POSTAL CULTURE

Writing and Reading Letters in Post-Unification Italy

GABRIELLA ROMANI

Postal Culture

Writing and Reading Letters in Post-Unification Italy

UNIVERSITY OF TORONTO PRESS
Toronto Buffalo London

Toronto Buffalo London
www.utppublishing.com

ISBN 978-1-4426-4708-4

Library and Archives Canada Cataloguing in Publication

Romani, Gabriella, author
Postal culture : writing and reading letters in post-unification Italy / Gabriella Romani.

(Toronto Italian studies)
Appendix includes letters transcribed from Italian newspapers.
Includes bibliographical references and index.
ISBN 978-1-4426-4708-4 (bound)

1. Communication and technology – Italy – History – 19th century. 2. Postal service – Italy – History – 19th century. 3. Written communication – Italy – History – 19th century. 4. Letter writing – Italy – History – 19th century. 5. Literature and society – Italy – History – 19th century. 6. Italy – Social life and customs – 19th century. 7. Italy – Civilization – 19th century. 8. Verga, Giovanni, 1840–1922. Storia di una capinera. 9. Serao, Matilde, 1856–1927 – Criticism and interpretation. 10. Epistolary fiction, Italian – History and criticism. I. Title. II. Series: Toronto Italian studies

P96.T422I83 2013 856′.809 C2013-904813-8

This book has been published with the assistance of the Charles and Joan Alberto Italian Studies Institute.

University of Toronto Press acknowledges the financial assistance to its publishing program of the Canada Council for the Arts and the Ontario Arts Council.

University of Toronto Press acknowledges the financial support of the Government of Canada through the Canada Book Fund for its publishing activities.

To the memory of Maria Giulia Faggioni, a beloved grandmother and daughter of the times here narrated

Contents

Acknowledgments

The first seeds for this project were sown many years ago during an independent study on epistolary writing in Europe when I was a graduate student at the University of Pennsylvania. The first acknowledgment is therefore due to Liliane Weissberg, who started me thinking about the tradition of epistolary fiction in Italy – a project that eventually developed into a study of postal culture.

There are many people I would like to thank for their role in making this publication possible. First of all, I am grateful to the anonymous readers for the University of Toronto Press for their feedback and criticism, which were very helpful for the revision of my manuscript and the fine tuning of my overall analysis of Italy's post-unification letter writing. Heartfelt thanks go to Ann Hallamore Caesar, who read and commented on the chapter on Matilde Serao, to Giuseppe Gazzola, who provided insightful feedback on the first two chapters, and to Rita Verdirame, who graciously guided me in gaining access to crucial material for my chapter on Verga. I am, finally, most grateful to my student Alyda Stabile for her invaluable help in translating and transcribing some of the original texts in Italian.

A note of thanks goes to Margaret Allen for helping shape the final product as it appears in the following pages, and a special one to Ron Schoeffel for all his extraordinary generosity throughout the editing process. He will be greatly missed.

Many people and institutions contributed to the realization of this project. Seton Hall University provided generous funding in the form of a summer research award and a sabbatical leave that allowed me to complete the manuscript. Several museums and libraries in Italy were instrumental in locating material – the illustrations, for instance – otherwise difficult to procure: Biblioteca Nazionale Centrale di Firenze,

Biblioteca di Storia Moderna e Contemporanea, Museo di Roma, Museo storico della comunicazione e Archivio storico di Poste italiane, and Istituto di Studi Storici Postali di Prato. In particular, Gabriella Aiello, Romolo Renzi, and Mauro De Palma went far beyond the call of duty to help me obtain images of postal calendars, which are a rarity even in archives. During my visit to the Museo storico della comunicazione, a little-known trove of Roman cultural and institutional life, Gabriella Aiello's and Romolo Renzi's guided tour was an unforgettable journey through a vast and striking collection of artifacts and inventions illustrating the history of communication in Italy. Finally, many thanks are owed to the staff of various libraries and archives who guided me through the challenges of the research process.

I owe my profoundest debt of gratitude to my family and friends, who sustained me throughout this time with love and encouragement. In particular, I want to acknowledge Davide Foa for providing me with much intellectual support and good humour. Julia, David, and Blue, thank you for your loving patience, especially Blue, who more than anyone shared with me many good and not so good moments in my study and must have wondered what could possibly be so engrossing on that computer screen to keep me from going out to play fetch with him on the field down the street.

A final note of acknowledgment goes to those friends who wrote me letters (a real rarity these days) during those long summers spent in the rural provinces of the Marche region when I was a child and teenager, and kept me connected to a network of friendship that I cherished immensely while separated from my normal life in Rome. I often went back to those memories as I reflected during my research on the power of letter writing to enhance a sense of wholeness and connection among correspondents.

PART I

1 Postal Culture after 1861: An Introduction

Una vita senza corrispondenze con i lontani mi sembrerebbe una mezza vita.

[To me, a life without corresponding with those who are far away would seem half a life.]

Emilia Toscanelli Peruzzi[1]

Ho detto che le lettere anonime producono nella società un monte di mali. Chi ne dubita non ha idea di quante se ne scriva. Dal camerino della portinaia al gabinetto del ministro, dalla soffitta della popolana al salotto della signora, allo studio dell'artista, all'ufficio del questore, del preside scolastico e del comandante del reggimento, fino al palazzo dell'arcivescovo e del sovrano, la lettera anonima va in ogni parte.

[I said that anonymous letters produce in society a great deal of trouble. Those who doubt it do not have any idea of how many of them get to be written. From the porter's quarters to the minister's office, the artist's studio, to the office of the police chief, school principal and head of the army, up to the palace of the archbishop and of the king, the anonymous letter goes everywhere.]

Edmondo De Amicis[2]

In this information age, when writing letters has become a rarity if not an oddity in a world of personal communication dominated by mobile communication, skype conversations, text messaging, and email exchanges, to engage in research on nineteenth-century epistolary practices may seem an anachronistic endeavour. And yet, the steady decline in our recent history of letter writing has only been matched by an increased academic interest in the topic, resulting in the publication of a large number of critical works focused on the history, practice, and fictional representations of letter writing. Paradoxically, we no longer write letters but love to write about them.

Drawing from the extensive scholarship available on this topic, this book examines the production in Italy of what critics have called "postal culture": the interconnections, influences, and overlapping of the literary representations, cultural productions, and communicative practices revolving around the letter – a medium of communication that, with the rise of nineteenth-century technologies and information systems (i.e., nationalized postal service, the electric telegraph, railway systems, and the printed media), stimulated a wider circulation of information and ideas.[3] As a result of technological advances and institutional reforms, by the second half of the nineteenth century the volume of letters exchanged nationally grew exponentially, as did the number of people who practised letter writing and who, most importantly for this study, joined in a network of new cultural and social activities. The metaphor of the network – understood as a connective system in the process of information exchange – exemplifies the important role played by the letter and letter writing as vehicles not only for the circulation of ideas but also in the formation of a collective national identity. As the Italian modern nation came into being with the political unification of the country in 1861, postal movements provided Italians with a new geography of national identity, based on the proliferation of post offices around Italy and on an enhanced sense of belonging to a group – the "horizontal comradeship" hypothesized by Benedict Anderson – of postal users eager to communicate for personal or professional reasons via letter with people living in distant locations (7). The construction of an imaginary presence is a fundamental feature of epistolary writing, because, when writing or reading a letter, interlocutors inevitably create an incorporeal presence of the people they are connecting with through the missive. On a national scale, such an incorporeal presence corresponded to the Italian body politic, the representation of a national identity viewed as an idealistic extension of the individual experience. Not surprisingly, officials of the Italian Post Office compared the new national postal system to a family with a set of norms "che regolavano il servizio di posta nelle diverse province italiane, felicemente ricongiunte in una sola famiglia" [that regulated the postal service in the various Italian provinces, happily joined together in a single family] (Paoloni 44).

The postal service created after the political unification of Italy was viewed by contemporaries as a main indicator of "progresso e civiltà" [progress and civilization] and a "fattore di unificazione" [an agent of unification], as Giovanni Barbavara di Gravellona, director of the Italian Postal Service, stated in his 1863 opening remarks for the first *Relazione statistica sul servizio postale* (Paoloni 43). In Italy as in other Western countries, nineteenth-century technological advances revolutionized the way people communicated and brought, along with the material improvement of the communication system, an awareness of new opportunities for connections with and exploration of the wider world. While the term "revolution" might not aptly describe this initial phase in the development of the Italian postal service, still today notorious for its chronic inefficiencies, the postal transformation of the second half of the nineteenth century represented for contemporaries a clear manifestation of modernity in the making. Indeed, the nationalization and modernization of the postal service, with its gradual expansion and vaster system of postal stations and routes disseminated throughout the nation, was presented by institutional functionaries and postal enthusiasts as a validation of national policies and of the impact that government interventions were having on people's daily lives. Emphatically, Giovanni Barbavara insisted that the task of the new Italian administration would not be complete until the postal service had reached "[i] più remoti villaggi del regno" [the most remote villages of the kingdom], where the delivery of mail would sometimes be the only concrete evidence of Italy as an institutional entity (Paoloni 49). In the decades following 1861, when the country was faced by questions of shifting national borders and undetermined national identities, the expanding national postal service and the concurrent epistolary practices provided Italians with a growing national consciousness, a new understanding of Italy as a national entity and of the opportunities, actual and imagined, that such a modern entity would unfold in the future.

Starting, therefore, with the premise that, after the political unification of Italy in 1861, institutional reforms and technological innovations sparked a significant growth in postal activities, this book explores the connections between the transformation of letter writing as a social practice and the growth of cultural productions in the epistolary format during the second half of the nineteenth century. As Armando Petrucci has pointed out, it was precisely in the Ottocento that "la corrispondenza scritta si trasformò da fenomeno sostanzialmente singolare, occasionale e in qualche modo controllabile, in un fenomeno socioculturale funzionale e strutturale rispetto allo sviluppo culturale ed economico della nuova società industriale" [written correspondence was transformed from a phenomenon substantially singular, occasional, and somehow controllable, into a sociocultural phenomenon that was structurally functional to the cultural and economic development of the industrial society] (133). Not only did more people engage in epistolary communication, but the letter appeared more frequently in the cultural domain, whether in the form of articles written as letters or in epistolary novels. The second half of the nineteenth century marked a turning point in the transformation of the print industry from a pre-modern, artisanal mode of production to a more advanced cultural industry. According to Fausto Colombo and Giovanni Ragone, it is possible to speak of a modern print media in Italy only after the unification of the country, and more specifically after the 1880s, when technological advances converged with the emergence of new cultural agents such as the professional writer, the figure of the publisher, and a rising consumer society and readership.[4] Letter writing as a form of communication and cultural production exemplifies this moment of transformation as the letter's material, social, cultural, and perceptual properties changed throughout the nineteenth-century in a similar fashion as in print culture: they both reflected and benefited from the modernization of the communicative system and the democratization of the cultural sphere, with a resulting wider educational and social access to reading and writing. Hence the letter's unique position as an interpretive tool for the analysis of Italy's post-unification cultural scene and literary discourse. As a means of connection in a complex of cultural and social intersections, the letter reveals the hidden dynamics and intricate interplay existing among the different interlocutors or cultural subjects (authors, critics, publishers, and readers) who inhabited and influenced a literary period still too often critically assessed through the exclusive study of single figures or ideas (Bourdieu 35). My study, although partly focused on the epistolary fiction produced by two specific late-nineteenth-century writers, Giovanni Verga and Matilde Serao, aims not so much at re-evaluating these authors' (traditionally neglected) epistolary productions within the context of their general oeuvre, but rather at discussing their works in the context of other forms of cultural production in the letter form developed in Italy during the post-unification period (including journalistic writings, critical reviews, correspondence, and educational publications). If it is true that "few areas more clearly demonstrate the heuristic efficacy of relational thinking than that of art and literature," then these epistolary texts, too, ought to be viewed as valuable instruments in an attempt to recover some of those elements (both in terms of form – genre, language – and content – specific themes and issues) that contributed to the formation of Italy's post-unification literary discourse (Bourdieu 29). Bourdieu's notion of the "field" is helpful here in configuring a method of analysis and a theoretical approach to the literary text, based on the principle that "no cultural product exists by itself, i.e. outside the relations of interdependence which link it to other products" (32). Such a synchronic approach expresses, furthermore, the need to bring further critical attention in literary studies to the heterogeneity and complexity of a literary period that, in spite of its deficiencies, contradictions, and ambiguities, undeniably marked a crucial moment in the formation of Italy's modern cultural identities. With this premise in mind, therefore, I proceed to formulate some general questions that have inspired, from the very beginning, my inquiry into the Italian nineteenth-century postal culture: Can the Italian epistolary fiction produced in the second half of the nineteenth century be considered merely the product of an earlier European epistolary tradition? Or should we interpret it in light of the developing national print media, as the result of a national convergence of factors, giving rise to a literature that responded to specific cultural, social, and literary demands in Italy? What new knowledge and understanding of this cultural period can be drawn out of what in Italian studies is usually viewed as a "minor" literary genre?

With the exception of two novels, Ugo Foscolo's *Ultime lettere di Jacopo Ortis* and Giovanni Verga's *Storia di una capinera*, nineteenth-century epistolary fiction in Italy has not attracted much critical attention. One reason for this neglect may stem from the fact that Italy, unlike other European countries, such as England and France, did not produce a large number of novels in the epistolary form. Even Foscolo's and Verga's epistolary novels have traditionally been considered "imperfect" works of art, as critics have often pointed to their inferior literary merit in comparison to the authors' other productions, respectively, poetry and fiction in a *verista* vein.[5] The real story of letter writing and modern Italy, however, is much more varied than can be inferred from the literary canon. Already in the seventeenth century various writers had adopted the epistolary form for their writing, among them Giuseppe-Antonio Costantini, who published *Lettere giocose, morali, scientifiche ed erudite* (1735), a collection of letters that enjoyed a large circulation for its time, and, in the second half of the eighteenth century, the Venetian

Abbot Pietro Chiari, who popularized this genre among the growing public of readers.[6] In the nineteenth century, then, with the emergence of postal culture, some of the most popular writers of the time, such as Giovanni Verga, Matilde Serao, Neera, Paolo Mantegazza, Angelo De Meis, and Marchesa Colombi, adopted it for their fiction.[7] Indeed, not only does late-nineteenth-century Italian fiction often feature the letter, but some of these authors became known to the larger public of readers thanks to their epistolary production.

Why did they write their fiction in the epistolary genre? Was their choice of the letter form merely a strategy to capitalize on a genre that had proven popular in Italy through the novels of Chiari, Foscolo, and other European writers? Had the epistolary genre at this point lost all its creative force? Or, if we ought to believe that the nineteenth century created, as critics seem to agree, "autentici monumenti epistolari" [authentic epistolary monuments] (Tellini 10), should we consider the epistolary fiction produced during this time part of the same cultural phenomenon that saw in the letter an unprecedented form of a more modern mode of communication? Can we look at the epistolary fiction produced in the late nineteenth century as a successful, rather than a failed, attempt to address some of the narrative and thematic questions underlying the authors' poetics and literary ambitions?

This book analyses the epistolary fiction produced by two authors, Giovanni Verga and Matilde Serao, in light of contemporary debates on specific thematic and artistic issues, and of the overlapping between these writers' publications in newspapers and books. My research draws mainly on documents produced during the authors' lives, – that is, reviews by contemporary critics rather than by later ones, correspondence exchanged with other intellectuals, and journalistic productions – with the aim of recapturing the historical and cultural atmosphere in which they produced their works and of revamping the critical debates surrounding these nineteenth-century literary texts, which have been lost or undervalued in the subsequent selective construction of the literary canon.

While Verga and Serao differ considerably from one another in terms of their overall poetics and literary production, it is possible to identify some common ground in the way they weave both innovative and traditional elements of the epistolary genre into their fiction. One innovative element may be found in their fictional representation of the letter, which, from being an emblem of solitary meditation, becomes a vehicle of public mediation; there is, in other words, a change of focus from individual order to social interaction, from the speaking "I" to the receiving "us," with a focus in the process of the communication on the desire to engage in interaction with one another. While historically the fictional letter has been interpreted as a strategy for delving into the characters' psychology or for creating a narrative of intimacy, the epistolary fiction of these authors emphasizes the actual process of communication, the delivery of the message contained in the letter novel. This shift from letter as confession, with the emphasis on the sender of the missive, to letter as communication, with the stress on the recipient of the message, suggests a rearrangement of agency in the epistolary discourse, one in which the reader of the letter becomes as important an agent as the writer. Reader-response criticism and reception studies have amply demonstrated the significant role readers play in the construction of textual meaning; what should be noted here, however, is the creation of a fiction in the late nineteenth century that favoured the communicative aspect of the epistolary exchange, the interplay between the correspondents inhabiting both the internal space of fictional interaction and the external arena of cultural production. Letter writing might be as ancient as writing itself, but it is in the modern age, to follow Foucault's insight, when the *raison d'état* transformed everyone into a subject of a modern state, that epistolary fiction becomes in the literary imagination an arena of public exchange and interaction.

A second element of commonality relates to a traditional aspect of the genre: the sentimentalism of these authors' epistolary fiction. The word "sentimental" appeared first in England in the middle of the 1700s and gained currency with the epistolary fiction of Samuel Richardson. While more recently "sentimental" has acquired a negative connotation of superficial emotionality, in the 1700s and 1800s it conveyed a more favourable meaning, one in which sentiments were equated to elevated feelings. This distinction between feelings and emotions is central to my reading of nineteenth-century Italian epistolary fiction, as I interpret the epistolary narratives of Verga and Serao in light of their rhetorical effectiveness, as an articulation of the language of sentiments, or "politica del cuore" [politics of the heart], as Serao put it, that was supposed to sensitize the growing readership of post-unification Italy to the issues or ideas presented in the fiction and, thus, shape public opinion. The language of sentiment, understood as a structured rhetorical tool, was presented as a harmoniously formulated set of values rather than an outpouring of irrational emotion.

"Cœur-responding," as David Henkin noted, is corresponding with the heart.[8] Because of the emphasis in epistolary writing on the naturalness of its language and style, critics have identified a feminization of the genre in the eighteenth century.[9] The eighteenth- and nineteenth-century epistolary tradition of Italy as well as of other European countries flourished on the assumption that women could best express their inner emotions in the letter form; and authors, both male and female, capitalized on this assumption and adopted the letter as a literary strategy for its communicative potential and rhetorical effectiveness. Furthermore, sentimentalism became in the late nineteenth century a popular narrative strategy used to infuse the text with strong didactic and entertaining elements. De Amicis's best-selling novel of 1886 was titled *Cuore* (*Heart*), a symbolic reminder that, as Leone Fortis, alias Doctor Veritas, once affirmed, "l'unità italiana è fatta nei cuori, e quando è fatta là, si fa presto a tradurla nelle leggi, nelle abitudini, nelle convizioni, nella vita" [Italian unification is made in the heart, and once it's there, it is easy to translate it into laws, habits, convictions, life].[10] Finally, while in the second half of the nineteenth century illiteracy rates in Italy were still very high, especially among women, female readers, particularly of the middle classes, had more free time to devote to leisure activities, including epistolary exchanges.

Women were seen by contemporaries as particularly talented in the art of writing letters; Paolo Mantegazza, for example, in his best-selling *Fisiologia della donna*, noted that "le donne emergono e brillano nello spirito e nel talento epistolare [...] L'uomo scrive in fretta perché ha altre occupazioni più serie [...] la donna invece ha quasi sempre meno da fare di noi" [Women are greatly gifted with the spirit and talent of epistolary writing ... men write in a rush, because they have more serious business to attend to ... women instead have almost always less to do than us] (194). From ancient times women had been thought of as natural writers of private letters. In ancient Greece, Alossa, the wife of Darius, was, according to Maria Luisa Doglio, (wrongly) identified as the first writer of a letter, and Cicero – a founding figure of the Western epistolary tradition – believed that Cornelia's letters to her son Caius represented the first example of letter writing in ancient Rome (i). Starting in the Middle Ages, furthermore, a tradition of female letter writers began to take shape: from Caterina da Siena, who contributed to the creation of a female tradition of letter writing, to fictional women such as Fiammetta, in *Elegia di Madonna Fiammetta* (c. 1343–4) by Giovanni

Boccaccio, who, adopting the Ovidian model of female letter writing, created a modality of fictional female writing that survived until modern times (Doglio i–vi). Finally, during the Renaissance, many women contributed to what was by then an established epistolary genre: the *lettera familiare*.[11]

As we move into modern European epistolary fiction, we find that women were often the main protagonists of epistolary novels. Elizabeth C. Goldsmith, in her edited volume on the French epistolary tradition, pointed out how this particular literary form in the eighteenth century was considered the best expressive form for the female voice (vii–xii). Both the English and French epistolary traditions flourished on the assumption that women could best express their inner emotions in the letter form. In fact, the most famous European epistolary novels (Richardson's *Pamela* and Rousseau's *La Nouvelle Héloise*) were centred on the love stories of female protagonists. A woman writing a letter was by the nineteenth century a common literary topos. Cécile Dauphin, however, cautioned against projecting these portrayals into the reality of the time and found that, while in nineteenth-century France epistolary literature thrived on narratives that presented women writing letters and epistolary manuals explicitly targeted a female readership, men were the main letter writers, and these manuals ultimately aimed at containing rather than encouraging female letter writing, which was viewed as a risky practice, potentially compromising for the good reputation of, especially, young ladies (122). Considering the very high rate of illiteracy among Italian women until the end of the nineteenth century, one can only assume that in Italy, too, men were the main letter writers, and yet, in Italy, too, the topos of the female letter writer emerged in popular culture. There was, therefore, a paradoxical distinction in epistolary discourse between a literary tradition of female letter writers and the reality of a world of correspondence dominated and supervised by men. The situation is further complicated when we find, as Laura Salsini has suggested in *Addressing the Letter: Italian Women Writers' Epistolary Fiction*, that several Italian women writers "adopted the letter novel – in all its variations – as a means to redefine both literary and social expectations of female experiences" (13). Modern epistolary texts by Italian women writers, especially in the twentieth-century, I would argue, became an instrument used to challenge gender and genre assumptions.

My research on late-nineteenth-century Italian postal culture acknowledges the scholarship so far produced on the genre and gender aspect of Italian epistolary fiction but focuses its attention on the historical contingency of this literary production, demonstrating that epistolary fiction was indeed intrinsic to the process of transformation of Italy's culture in the post-unification period and that real correspondence and epistolary fiction were not only related but intrinsically intertwined, as they shared a similar set of conventions and practices. As Amanda Gilroy and W.M. Verhoeven aptly put it, the epistolary form "is historically and culturally specific: the letters that come to us, and the fiction/histories that we write about them, rely on and bear the traces of particular historical practices" (1). In light of this interpretation, I proceed to explore the "envelope of contingency," to use Mary Favret's definition – that is, the social, historical, and cultural environments surrounding the creation of literary texts, without which an analysis of the literary productions of that period would be essentially incomplete (56).

Critics have traditionally focused on a diachronic study of letter writing, emphasizing its historically contingent nature –"un complemento all'educazione intellettuale che procede di pari passo con la disciplina dei costumi" [a complement to the intellectual education which proceeds hand in hand with customs] (Zarri xvii) – but not enough attention has been paid to the relationship existing between real letters and fiction, between the daily practice of writing letters and the imaginative production of novels or short stories in letter form. Late-nineteenth-century epistolary fiction, like other cultural and social discourses of the time, reflected the transformation of the burgeoning cultural industry, which, through the influence of new hegemonic forces, marketing policies, and ideological imperatives, was forging new paradigms of cultural production based on a more flexible interrelation between literary and non-literary forms of communication (Ragone 3).

Cultural historians and literary critics have recently approached epistolary fiction through the analysis of its historical and cultural specificity. Amanda Gilroy and W.M. Verhoeven claim, for instance, that "the epistolary form, as we have come to recognize it, is historically and culturally specific" (1); and, in a shift from production to reception of the text, Roger Chartier emphasizes in *History of Reading in the West* that "we should keep in mind that no text exists outside of the physical support that offers it for reading (or hearing) or outside of the circumstance in which it is read (or heard). Authors do not write books: they write texts that become written objects [...] All these objects are handled, in various ways, by flesh and blood readers whose reading habits vary with time, place and milieu" (5). In this vein, in a study on Mme de Sévigné's letters, Roland Racevskis demonstrates how the development of the French postal system during the reign of Louis XIV in the second half of the seventeenth century notably shaped the "patterns of everyday activity," a change that was then reflected in the literary production of the time (51). And, finally, for Thomas O. Beebee, who has written extensively on letter writing, the letter represents "a Protean form which crystallized social relationships in a variety of ways," one that could enter the fictional world only once it had "appropriated the status and power the letter had acquired from its established functions from other discursive practices" (*Epistolary Fiction in Europe* 3).

According to these interpretations, epistolary representations in fiction would be impossible without the existence a priori of a system of communication that would legitimize the very concept of authenticity upon which much epistolary fiction is based. But Beebee advances even further this notion of interconnection between the literary and non-literary uses of letter writing, and, relying on Richardson's and Derrida's formulations on the relationship between letters and literature, points out how difficult it is to "extricate any single letter from a nexus of texts that include both natural and fictive discourses" ("Publicity, Privacy" 67). In sum, "fictitious novels followed the same conventions and satisfied the same expectations as did real letters" (ibid. 65). This is not to say that readers of fictional letters simply interpreted the fictional text in light of their personal experience with correspondence and postal service, or that authors of epistolary fiction merely incorporated the letter form to entice the reader into a process of obvious identification, but rather that both readers and authors of epistolary fiction responded to and functioned in a world of shared practices and perceptions. Precisely because of the historical and cultural interrelation between customary practices of postal communication and fictional letter writing, authors and readers alike have relied on the overlapping of these two worlds – of real postal usage and the fictional representation of it.

People have been writing letters since time immemorial, but it is only from the late nineteenth century, when a nationalized postal service was created, that one could speak of a modern system of communication. Until the political unification of Italy in 1861, postal communication was organized locally, at the level of each individual state, which had its own postal system with its own tariffs and sets of rules. There were still, around the 1850s, three main areas of mail distribution in Italy: a north-western area, headed by the Kingdom of Sardinia; a north-eastern area, controlled basically by the Austro-Hungarian Empire; and a southern one, connected to the Kingdom of the Two Sicilies, where the delivery of mail was limited to the main cities. On 15 December 1860, a *Regio decreto* established Italy's first nationally organized postal service, which entailed the extension of the Piedmontese postal system to the rest of the Italian territory. And on 5 May 1862, the first comprehensive law relating to the national postal service recognized, after long debate, the monopolistic principle according to which postal communication was a public service, provided and regulated exclusively by the government (Paoloni 47).

The relatively more efficient service and a more simplified system of postal rules and lower costs prompted a significant increase in the volume of postal activity even in the first year of postal reform; a sign, according to Giovanni Barbavara, "della crescente prosperità del paese nostro e della cresciuta attività sociale che sotto libero reggimento riceve più rigoroso impulso" [of the increased prosperity of our country and of the growing social activity which under a free rule receives a stronger impulse] (quoted in Paoloni 48). In addition, the social transformation of the country, as a result of its relative urbanization and new class stratifications, along with national educational efforts, created a larger than ever group of postal users. But this quantitative growth also signalled a significant change in how communication itself was perceived in society. The creation of a network of postal users, with its multiple trajectories of possible national and international communication, transformed people's perception of space and notions of personal as well as collective boundaries. All over Europe, cultural and political manifestations of shifting national identities (growing imperialistic ambitions, larger-scale tourism, and international cultural events such as world fairs, to name a few) were expressions of a similar phenomenon. Suddenly the world appeared smaller and more accessible. People travelled more often, and not only by way of actual physical movement, as popular illustrated magazines and books offered imaginary journeys in exotic and far-away lands to the ever-growing consumer-minded readership. As the world became more accessible, the literary imagination increased its range, offering a tempting array of narrated journeys, which on the one hand enabled readers to visit places still unknown to them but on the other confirmed the static reality of their daily existence (Brilli 9). The popularity of authors such as Emilio Salgari and Edmondo De Amicis in the late nineteenth century attests to this phenomenon. When De Amicis, for instance, wrote in the 1870s about Spain, Morocco, Holland, or the Exposition in Paris – articles published in *La Nazione* and *L'Illustrazione Italiana*, often in the form of letters – he struck a chord with the reading public of the time because of his capacity to satisfy readers' curiosity about the world and at the same time allow them to remain in the comfort zone of their familiar space. Stylistically and functionally, the letter played no small role in an author's attempt to bridge the distance, both geographical and cultural, between the private world of the reader and the public landscape depicted in the travel narrative. Since the letter straddles the gulf between presence and absence, it leaves the interlocutors in an ambiguous intermediary space where they are neither totally united nor totally separated (Altman 43). This mediatory role of epistolary writing is what ultimately creates communication, whether in imaginative or practical terms.

In the 1870s, writing letters and sending postcards to friends and family members living far away became a common practice. As was noted in a popular nineteenth-century epistolary manual, "Chi di noi non ha parenti e amici in altre città d'Italia, se pure non all'estero, e anche fuori d'Europa, e non tiene con loro nutrito commercio epistolare?" [Who among us does not have relatives and friends in other cities of Italy, if not abroad, and also outside of Europe, and does not keep a frequent epistolary exchange?] (Gelli x). That Italy's late-nineteenth-century postal culture was vibrant and growing is also demonstrated by the unprecedented number of publications written about or related to epistolary writing. Manuals on how to write letters became best-selling books – a phenomenon belonging to the popular trend of "how-to" publications used by "quella maggioranza anonima che si serviva della scrittura in modo strumentale e suppliva alla propria limitata cultura consultando grammatiche, dizionari, manuali epistolari" [that anonymous majority of people who used writing instrumentally to make up for its limited culture by consulting grammars, dictionaries, and epistolary manuals] (Antonelli, Palermo, et al. 9). Some of these books were manuals used in class as textbooks for instruction in rhetoric and aesthetics; others, the more commercially oriented, were marketed to the general public and, having no theoretical pretensions, simply provided readers with a compilation of conventions and rules to be applied to correspondence of both a personal and a commercial nature. Moreover, letters appeared with frequency in newspapers, either as part of featured columns, referred to commonly as *Piccola posta* or *Corrispondenze* or *Conversazioni*, in which publishers and journalists kept up an ongoing dialogue with the reading public, or as *lettera aperta*, op-ed articles written in letter form by well-known public figures and addressing readers on issues of current interest. Finally, the letter became in the late nineteenth century a literary strategy that several writers adopted for their novels and short stories.

The following chapters address different aspects of letter writing as it evolved from the 1860s onward. Chapter 2 is devoted to the emergence of postal culture in late-nineteenth-century Italy through an analysis of the history of the Italian postal service and of the letter as it appeared in the growing and evolving print media of the time. In particular, I look at how letters published in newspapers or presented in epistolary manuals constituted both instruments and examples of the larger circulation of ideas that occurred in the post-unification period. Chapter 3 is devoted to Giovanni Verga's *Storia di una capinera*, the novel which, with a remarkable sale for that time of 20,000 copies, brought the Sicilian writer to the attention of the national public of readers. Based on a popular theme in epistolary fiction – a young woman forced into a nunnery by the family – this novel's success was closely related to the author's ability to encapsulate in his narrative some of the main cultural discourses taking place at the time. His representation of human emotions and, in particular, of the "misteri del cuore" [mysteries of the heart], is the departure point for my analysis of contemporary cultural debates on "love" and the novel. Chapter 4 analyses the sentimental epistolary fiction produced by Matilde Serao, a journalist and novelist who infused her fiction with elements taken from her regular correspondence with her readers. The "sentimental" aspect of her fiction is interpreted in my analysis as part of the author's literary effort to create a narrative that could, on the one hand, represent her female readers' life experiences and, on the other, engage them in meaningful discussions on topics the author deemed relevant to them but also to the current literary discourse. I outline a fundamental distinction between sentiments and emotions that challenges the way in which Serao's sentimental fiction has traditionally been interpreted. Lastly, the book offers in the Appendix the full text of several letters, partially quoted in earlier chapters, that were originally published in nineteenth-century newspapers and that, to my knowledge, have never been reprinted in their original complete newspaper version. These archival documents reveal the richness of the cultural debates that informed the cultural and literary discourses of post-unification Italy.

The nineteenth century may indeed be considered the postal century par excellence, marking the highest point in the history of letter production and consumption, a practice that until then had been the prerogative of an elite and narrow section of the population and that afterwards began to gradually wane as a result of the introduction of new technologies in telecommunications – the telephone, the Internet, and electronic mail – which fatally pronounced the "death of letters," and, soon, of the Post Office – at least as we have known it for the last two hundred years – both casualties of our cyber-culture.[12] Far from being part of a dead literary tradition, however, epistolary novels continue to be written and to be read – testifying to the resilience of a genre that has proved to be remarkably adaptable to the changes we have adopted in the production and circulation of ideas in our modern times.

2 Writing and Reading Letters: The Nationalization of the Italian Postal Service, Epistolary Manuals, and the Print Media

For those who were born before the Internet era, the sheer speed of circulation of news, not only public but personal, is at times confounding: as a result of technological developments, much has changed in the past few decades in the way we receive and produce information. This reaction is not without precedent: despite the industrial backwardness of nineteenth-century Italy, people must have felt startled by the changes in the modes and timeliness of communication when, as a result of institutional interventions and technological developments (such as the construction of a more extensive railway system and the invention of the electric telegraph), the circulation of news became significantly faster in the post-unification period. Of course, complaints about the inefficiency of the Italian Post Office may be found at any time in its history, and especially at its inception when the distribution of national postal services was still sparse, but one should not underestimate the impact that the creation of a unified and centralized postal system had on people's lives as well as on the general perception in Italy that the country had finally embarked on a journey towards modernization.[1] As David Henkin points out in a study on nineteenth-century postal history, "Seen from the perspective of ordinary people, what may have been most remarkable about the middle decades of the nineteenth century was the novel experience of being accessed and addressed by a system of mass communication" (x). This chapter seeks to investigate the cultural and social implications of such novelty and the way in which the letter – whether in the form of personal communication or epistolary manual or newspaper piece – coalesced with other cultural and social trends in advancing a circulation of information that was unprecedented and that gradually became wider and more inclusive, crossing gender and class divisions and sparking new forms of actual but also imaginative expression of written communication.

The Nationalization of the Postal Service

While the postal service had existed in Italy since Roman times, one may speak of a modern system of postal communication only after the 1860s. If the term "revolution," used by some critics in reference to the transformation of the postal service at that time, might seem an overstatement, infused with a somewhat romanticized vision of the new technology's prowess – especially in a country like Italy, where industrialization was lagging in comparison to other European countries – its recurrence in the literature relating to the nineteenth-century postal system certainly attests to the effect that such innovations had on Italian society, both in practical terms and, at a more abstract level, in the collective perception of Italy as a country moving forward with the rest of Europe on the path of progress.[2] At this historical turning point in the technological transformation of the modes and means of communication, the letter became a main carrier – symbolic and literal – of news and ideas among the growing number of postal users. As a contemporary academician and publicist, Emilio Morpugo, wrote in 1883:

> E' così intima l'attinenza fra le relazioni epistolari e il grado di sviluppo delle condizioni sociali, e specialmente negli ultimi quarant'anni si son viste moltiplicare così trabocchevolmente le corrispondenze postali in tutto il mondo civile, anche fra persone del più umile stato, che la narrazione dei progressi e delle riforme del servizio della posta gareggia oggigiorno d'importanza con la storia delle maggiori scoperte. (3)
>
> [The connection between epistolary relations and level of progress has been especially strong in the last forty years; the civil world has seen its postal correspondence multiply so considerably, even among people of humble origins, that the narration of the progress and reforms of the postal service competes today for importance with the history of major discoveries.]

Only a few decades earlier, people spoke of the postal services in Italy in much less enthusiastic terms. Alessandro Manzoni, to mention a well-known figure among nineteenth-century postal users, complained frequently in his letters about the unreliability of the post. In his correspondence with Claude Fauriel, for instance, he lamented that the terrible slowness of the mail service resulted in frequent miscommunication among letter writers. In a letter to Fauriel, dated 13 July 1816, he mentioned that he did not understand his friend's words, and assumed that a previous letter from Fauriel had probably been lost. Without hesitation Manzoni blamed the post office for this: "On ne risque jamais de faire un jugement téméraire en rejetant la faute sur la poste: aussi j'étais très persuadé que quelque lettre de vous s'y était égarée" [One never risks being guilty of a rush to judgment by blaming the post office: I was therefore quite sure that some of your letters had been lost].[3] But even when the correspondence became more regular between the two friends, Manzoni advised Fauriel to be wary of the postal service: "Tout fier encore de votre lettre, il me semble presque que nous sommes en correspondence réglée, mais une triste expérience m'avertit que ce bonheur est une exception. Cette même expérience me dit qu'il n'y a absolument rien à compter sur la poste: ainsi veuillez guetter les occasions" [Still quite proud about your letter, it almost seems like we are having a regular correspondence, but unfortunate experience tells me that this happiness is an exception. The same experience tells me that one cannot count at all on the postal service: therefore please be on the lookout for other means].[4] While not trusting the postal service for the delivery of personal mail, he thought that commercial mail was more reliable. In a letter of 1821, Manzoni wrote, "Henriette n'a pas reçu une seule lettre de sa cousine, qui cependant lui a écrit plusieurs fois ainsi que nous l'avons appris d'une autre lettre qui lui est parvenue, peut-être parce que c'était une lettre d'affaires, car celles-là arrivent beaucoup plus régulièrement" [Henrietta did not receive a single letter from her cousin, who apparently wrote many times according to another letter she received from her, maybe because it was a business letter, since they arrive more regularly].[5] If business correspondence was delivered more regularly, business people did not hesitate to complain about the daunting task of delivering books and other items of print media, such as periodicals, across state borders on the Italian peninsula, using the postal service. On 20 July 1822, Giovanni Pietro Vieusseux, an important cultural figure in Florence, wrote a letter of complaint to the general director of the Tuscan Postal Service:

> Mi prendo la libertà di rinnovarle le diligenze che tante volte ho avuto occasione di farle, riguardo alla tassa eccessiva alla quale sono state fino adesso sottoposte nel Regno Lombardo Veneto i libri ed in generale

> qualsiasi foglio stampato che dalla Toscana vogliasi spedire colà col mezzo della Posta, doglianze che potrei fare a nome di tutti i librai, stampatori, autori, editori ed amatori di libri in Toscana ben sicuro di non essere da loro smentito. (Carpi 90)
>
> [I take the liberty to renew the solicitations that I have already had the occasion to express, in relation to the excessive taxation imposed in the Lombardy-Veneto Kingdom and on any printed sheet that is sent there from Tuscany by way of postal service. This is a complaint that I could make in the name of all booksellers, printers, authors, publishers and lovers of books in Tuscany, as I am more than sure not to be contradicted by them.]

Vieusseux was certainly not alone in his preoccupation with how to reconcile the needs of Italy's growing publishing industry with the country's still fragmented and inefficient postal system; but with the nationalization of the postal service in 1861 the increase in the amount of postal activity attests, if not to a sense of Italians' full confidence in the services provided by the *Poste italiane*, certainly to their more frequent use of it.

Both Manzoni and Vieusseux relied on a postal system that, until the political unification of the country, was regulated according to the rules of each individual state. A letter sent from Turin to Naples had either to face the often treacherous and uneven conditions of the roads and limited railway systems existing in the various states or rely on a maritime postal service run by foreign companies (mainly English and French). Most importantly, it had to deal with the variety of tariffs as well as the different currencies and procedures of the various mailing systems, complications that inevitably resulted in delays and therefore complaints by both private correspondents and professionals such as publishers, bookstore owners, and newspaper editors.[6] Joseph Luzzi has rightly argued that one of the impediments to national unification was the fragmented reality and "semiotic volatility of the various Italian currencies" (39). "La questione della lira" [the question of the lira], as Luzzi put it, was as dividing a factor as "la questione della lingua" [the question of the language] and, if the difficulties of mail circulation never reached the urgency of a true *querelle* in the political and cultural discourse of the Risorgimento, the need to normalize the postal system became a priority in the immediate post-unification period, when the new government decided to pass a series of postal decrees that became effective immediately after the official proclamation of the Kingdom of

Italy on 17 March 1861.[7] While addressing obvious practical concerns of national communication and the need to centralize and reorganize the local postal services, the nationalization of the postal system served also as the government's attempt to showcase its readiness to tackle national questions and thus offer a first tangible indication of the steps the country was taking to achieve national unity and progress. It was as much a rhetorical as a material response to the need to provide concrete examples of the benefits of unification. Writing and receiving mail thus gained symbolic value, and it was presented teleologically as a clear marker in the civilizing process of society – understood both in moral and material terms. As Morpugo put it, "Si può dire certamente nello stesso modo, a proposito di un movimento postale più o meno sviluppato: il grado a cui è pervenuta questa forma di pensiero è l'indice sicuro del grado di civiltà a cui il popolo ha potuto elevarsi" [One can surely say the same regarding a postal movement more or less developed: the degree to which this form of thought exists indicates the degree of civilization reached by its people] (8). Such improvement was mainly promoted in educational and cultural terms. The circulation of news was perceived as a formidable tool for creating and educating the nation, for distributing knowledge – albeit according to specific hierarchical orders of class and gender distinction – thus promoting a sense of belonging to the collective national community, and reducing distances among individuals – both in the abstract and in practical ways – by way of communication.

Before Italy's political unification each state had a postal service that depended mainly on local variables, such as the state's currency, physical infrastructure, and human resources (number of post offices and postmen), as well as on bilateral agreements for postal transportation and delivery between individual states of Italy and international nations. The most advanced postal service was offered by the Kingdom of Piedmont and Sardinia, which relied on a strongly bureaucratic and centralized system of postal distribution and on its leading position within Italy in international postal traffic, as it controlled all postal exchanges between the Italian territory and France, Switzerland, Spain, England, and Belgium; by contrast, both the Papal State and the Kingdom of the Two Sicilies were noted for their limited postal distribution. The Papal State, never eager to embrace innovations, carried out a first postal reform in 1826 (compared to the 1818 of the Kingdom of Sardinia), and it was only with Pius IX, in the late 1840s, that the postal system went through a meaningful administrative transformation. As for the Kingdom of the Two Sicilies, although Ferdinand I reorganized the postal system between 1815 and 1819 by instituting a more centralized administration of postal services, the territory of southern Italy continued to lack post offices, and postal service outside the main cities remained basically sparse and unreliable.[8] North-east Italy (the Lombardo-Veneto area) was under the Austrian-Hungarian postal jurisdiction and offered a relatively efficient service compared to other parts of Italy. The stamp was introduced in Italy for the first time in 1850 (Paoloni 16).

A change to this fragmented situation came with the first national postal reform introduced on 15 December 1860, in the form of an extension of the Piedmontese postal system to the annexed Italian territory. This initial legislative intervention, as Paoloni points out, reflected more a formal administrative integration of previously existing postal systems than a real reorganization of the national postal service, which remained tied for years to old customs and local (dis)services (46). On 5 May 1862, however, an important comprehensive postal law was passed, which introduced more uniform national standards of postal tariffs and practices, among them a fixed tariff of fifteen cents for letters weighing up to ten grams, regardless of the distance to their destination.[9]

The annual reports written by the general director of the Italian Post Office – under the jurisdiction of the Ministry of Public Works until 1889, when it was turned into a ministry of its own (Ministero delle Poste e dei Telegrafi)[10] – show that within a few decades the volume of correspondence processed by the Italian postal service increased considerably. In 1863, 71,543,346 letters were mailed in Italy; by 1873 the figure had reached 104,502,431.[11] Ten years later, in 1883, the volume of mail had increased by 50 per cent: 156,684,082 letters were sent, with Rome, Milan, Florence, Turin, and Naples as the busiest centres of postal activity. Considering the reality of a postal service that had just been nationalized and the fact that the country still had very high illiteracy rates, these figures represent a remarkable level of letter-writing activity. Although non-literate people also sent and received letters – with letters being dictated by and read to people who could not read and write – these numbers show that, whether literate or illiterate, people relied more and more on letter writing for their private or commercial correspondence. Writing, mailing, and receiving letters became a common practice of daily life.

But who wrote these letters? Who were the typical letter writers of late nineteenth-century Italy? Certainly, many of the letters included in the figures provided above were of a commercial and official nature. Just as many, though, belonged to private exchanges, as attested by the flourishing of nineteenth-century correspondence, much of which has since been made available in print.[12] It has been suggested that people from the rising middle classes and living in urban centres, where the literacy rates were highest, made use of this medium most frequently.[13] The nineteenth century in Italy also saw the emergence and affirmation of the so-called "lettera borghese" [bourgeois letter], understood in its wide social connotation as a communicative tool used by middle-class correspondents who were able to read and write with varying levels of competence, from literate to semi-literate (Petrucci 139). No longer an exclusive means of communication for the literati or the upper classes, the letter began to be used more frequently by ordinary people, not only by people such as shop owners, clerks, artisans, teachers, and housewives – the "ceto di frontiera" [frontier class] to use an expression coined by Paolo Macry for the expanding lower middle class of liberal Italy – but by people of all social backgrounds who, under new life circumstances, for the first time needed to communicate by post.[14] Letter writing became, as Roger Chartier points out, "a sort of 'ordinary' everyday and private writing, like the accounts book, the recipe book or the family record book" ("Introduction" 2). It is instructive, in this regard, to look at a project of recent digitization of nineteenth-century letters (*CEOD: Corpus Epistolare Ottocentesco Digitale*), which included 1,300 letters by seventy-three letter-writers, some of whom, as the authors of this research project point out, were lower-class individuals who, though remaining "ignoti alla storia con la S maiuscola (o da essa solo sfiorati)" [unknown to history with a capital H (or only marginally touched by it)], regularly used the mail for various types of communication (Antonelli et al. 9).[15] The Longhi family of Casalzuigno (a small town near Varese), for instance, left a collection of letters exchanged at the end of the nineteenth century when several members of this peasant family left Italy for Switzerland, France, and South America in search of better living conditions. While these letters lack the linguistic and stylistic sophistication of contemporary epistles written by more highly educated correspondents, they nevertheless constitute an eloquent example of the readiness with which ordinary families used the letter – the only medium, indeed, that enabled people like the Longhis to maintain contact with family members in other parts of the world.[16]

Studies of letter writing as a practice used across the social spectrum illustrate how, in spite of the still pervasive low rates of literacy – at the beginning of the 1870s about 70 per cent of the national population was still illiterate – not only literate people read and wrote letters.[17] Emilio

Franzina, for instance, has convincingly argued that letter exchanges were not the exclusive prerogative of members of the upper and middle classes. Correspondents from the lower classes also contributed to a considerable body of letters, often imbued with "temi fissi, di formule, di schematizzazioni e di frasi fatte, spesso con andamento ritmico o proverbiale" [pre-established themes, formulas, preconceived schemes and sentences, often with a rhythmical and proverbial tone], but which, according to Franzina, were symptomatic of the growing national and international mobility of Italians during the second half of the nineteenth century (27). Preservation of these letters has certainly been a problem for scholars in this field. Archival collections of letters exchanged among the lower classes are often sparse and fragmentary, for letters produced in these social and economic environments were more likely to be discarded or lost, but the lack of such documentary evidence is partly compensated by literature. Several authors of the nineteenth and early twenthieth centuries, from Manzoni to Pirandello, have provided examples of epistolary exchanges by illiterate correspondents, such as peasants or immigrants, who, though unfamiliar with the written language, found themselves needing to write a letter.[18] A most famous portrayal of correspondence among Italian peasants may be found in Alessandro Manzoni's *The Betrothed* (1840). Although the story of Renzo and Lucia takes place in seventeenth-century Lombardy, Manzoni portrays the protagonists' letter-exchange as if it is occurring in the author's own time; not much had changed in the last two hundred years, he laments, in the way peasants conducted epistolary communication. "Ma per avere un'idea di quel carteggio, bisogna sapere un poco come andassero allora tali cose, anzi come vadano," Manzoni warns, "perché, in questo particolare, credo che ci sia poco o nulla di cambiato" (1173) [To have an idea of what this correspondence was like, we need to know how that sort of business was transacted in those days – or rather how it is still transacted today, for we doubt if there has been much change].[19] Renzo, "che si struggeva di mandar le sue nuove alle donne, e d'aver le loro" (1173) [how impatient he was to send news of himself to Lucia and Agnese, and to hear news of them, 495], though able, with difficulty, to read, cannot write; if he wants to communicate with Lucia, he has to ask for the help of a third person, who writes and delivers his letter to Agnese. Lucia's mother, herself illiterate, asks her cousin Alessio to read Renzo's letter and write her response to him. Communication among the different parties, Renzo, Agnese, and Lucia, eventually takes place, but it remains miraculously and precariously dependent on an external figure, the *scrivano* [street scribe], who mediates and, at times, heavily emends the correspondence (see figure 1). The *scrivano*, a literate family member or acquaintance, or, sometimes, the postman, was commonly the individual on whom people who could not write had to rely for communication by post. But such reliance came at a price, for "chi ne sa più degli altri," as Manzoni notes, "non vuole essere strumento materiale nelle loro mani; e quando entra negli affari altrui, vuol anche fargli andare un po' a modo suo" (1173) [a man who knows more than his neighbours does not care to be a passive tool in their hands, and once he has become involved in their affairs, wants to give them a little guidance, 497]. Even in the best of circumstances, these types of mediation resulted in the correspondents' loss of a most essential property of letter writing: the intimacy and privacy of the familiar letter. A sense of violation and disempowerment unavoidably pervades mediated epistolary communications. Maria Messina, in a short story titled "Nonna Lidda," describes well the frustration felt by a mother, being helped by the cook to write a letter to her son in America, when she is unable to stop the cook from meddling with her correspondence:

> Ma la gna' Lidda aspettava con premura grande, e se il postino tardava d'un giorno, si disperava. Mentre il coco le scriveva la risposta, essa lo stava a guardare coi suoi occhietti verdolini, lo guardava nella mano, nella penna sottile che scriveva le parole dettate. Ma scriveva proprio come lei dettava, mastro Nitto? No, un giorno glielo aveva detto: -Mica si scrive come si parla! Ma però si capisce lo stesso.
>
> Da allora non ebbe più pace la gna' Lidda. [...] E restava a guardare fin che il coco chiudeva la sopraccarta gialla coll'indirizzo stampato – gliela mandava ogni volta il figlio – e impostandola le restava sempre il dubbio che il coco non avesse scritto quel che lei aveva dettato. (127)

> [But *gna'* Lidda waited with a sense of great urgency, and if the postman was a day late, she would despair. While the cook wrote her responses, she would watch him with her small green eyes, watch his hand and delicate pen that transcribed the words she dictated. But was *mastro* Nitto really writing down exactly what she was saying? No. One day he told her: "You don't write the way you talk! People still understand."
>
> From that moment on, *gna'* Lidda no longer had any peace. [...] And she would stay there watching until the cook closed the yellow envelope with the printed address – her son would always send it to her – and mailing it she always had the doubt that he hadn't written exactly what she'd told him to write.][20]

Figure 1: Francesco Coleman, *Lo scrivano pubblico* (1880). Watercolor. Museo di Roma in Trastevere, Italy.

Italian immigrants who left their country in the nineteenth century for South and North America were more likely than other Italians of similar social background to write letters, in an attempt to stay in touch with family members left behind in Italy. Scholars of immigrants' letters, a burgeoning field of study, point to the fact that nineteenth- and early-twentieth-century Italian migration played a fundamental role in spreading literacy among the lower classes (Gibelli and Caffarena 563); such correspondence, sparse and fragmentary though it might be, eloquently demonstrates that by the 1880s postal movements had gone beyond national borders to take on transnational trajectories. Letter writing enabled Italians not only to maintain contact with friends and family members but also to participate in a network of information exchange, the nineteenth-century *postal culture,* which exposed them to new ideas and customs, or simply unfamiliar realities that would otherwise remain unknown to them. Immigrants described in detail their new lives in a foreign land, and the new realities of their social and economic condition, sometimes omitting or embellishing elements of their stories. These narratives were indeed what provided many Italians with the impetus to emigrate and join a relative or fellow countryman in a foreign country with the hope of finding better working or living conditions. Whether carrying good or bad news, letters eroded distances and enhanced a sense of connectedness that defied local isolation and traditional norms of social order and mobility, ultimately favouring the creation of new "imagined spaces" that translated into the construction of new subjectivities (How 1–3) within both the social and cultural domains of Italy. As Bruce Redford points out, personal letters do not simply mirror worlds, they create a world of their own.[21] For Italians – whether literate or illiterate, whether living in isolated, rural areas or urban settings – being exposed to letter writing meant becoming part of a world much larger than the one they inhabited.

In the immediate post-unification period, the most typical item of postal mailing was the letter, although other forms of information exchange circulated commonly by mail.[22] Newspapers were delivered by postal service at the relatively low cost of one cent. The newly formed government understood the importance of cultural interventions and tried to capitalize on the growth of the private print industry for its educational efforts by lowering the price of mailed printed material, in line with similar postal policies introduced by other European countries. In the very first report published by the director of the Italian Postal Service in 1863, Giovanni Barbavara pointed out how the low newspaper tariff (one of the lowest in Europe with the exception of Belgium and Holland) facilitated the circulation of ideas and the creation of a common Italian identity. "La riduzione della tassa dei giornali a un centesimo per esemplare," he wrote, "ha largamente favorito la diffusione delle pubblicazioni periodiche, che tanto giovano all'educazione politica del paese" [The tax reduction to one cent for newspapers has largely favoured the circulation of periodicals, which so much improve the political education of the country] (*Prima relazione sul servizio postale* 18).

Among the various postal initiatives connected to this project of national education was the *calendario postale* [postal calendar], an illustrated calendar, normally given as a Christmas gift to postal patrons by postmen. Widely circulated across the national territory, postal calendars were hung in homes and also displayed on town walls during religious celebrations or rural fairs.[23] Following the tradition of the almanacs and moon-phases calendars ("lunari"), which since the Renaissance had been widely circulating, thanks to the figure of the "venditore ambulante" [street vendor], postal calendars featured illustrations that commonly displayed a didactic message. The visual as well as decorative aspect of the calendars was, as in all similar printed material, a crucial element of the popularity of a genre coveted for practical reasons – keeping track of time – but also for aesthetic ones – to decorate an otherwise bare wall or simply to demonstrate through the ownership of printed artefacts a certain social status or level of know-how. While limited in terms of variety – within a range of repetitive and predictable iconography – these illustrations changed from year to year according to what was perceived as being historically and culturally relevant at the moment. The 1871 *calendario postale*, for instance, featured a *bersagliere* (soldier) standing next to a young peasant Roman girl (fetching water at a fountain), celebrating thus the recent take-over of Rome and the completion of a political unification still an unfamiliar novelty in the nation as a whole (see figure 2). The 1898 postal calendar, by contrast, presented on the front cover a dashing young postman surrounded by an art-deco floral decoration, while in the back, a collage of images portrayed different archaeological sites, from the Roman Forum to the Temple of Edfu in Egypt, which on the one hand reflected the growing phenomenon of mass tourism and curiosity about distant and exotic places and on the other expressed the sense of national pride in an artistic and archaeological patrimony that rivalled in beauty and richness those of other ancient civilizations (see figure 3). Coloured by the mood of the time, graphically and thematically shaped by the technological improvements and cultural and ideological mandates of the moment, these postal calendars – along with almanacs and illustrated religious material – were part of a burgeoning popular culture in Italy that encouraged a sense of belonging to the Italian nation through the production and consumption of artefacts with a wide social appeal. Materially and temporally ephemeral though they were (the reason why very few calendars have been preserved for posterity), postal calendars illustrate how postal printed material not only captured the spirit of the time but also served the national educational program by reaching corners of the national territory and sections of the population where both schoolbooks and other forms of printed material had yet to arrive. With little writing, apart from the names of months, calendars relied on their images to communicate their narratives of postal promotion by projecting a notion of postal service, and consequently postal users, that appeared advanced (in including,

Figure 2: *Calendario Postale 1871*. Archivio Storico di Poste Italiane, Rome, Italy.

for instance, symbols of technological development such as trains, telegraph lines, and faster vessels), modern, and worldly.

The idea of social and cultural advancement through postal communication reverberated strongly in contemporary publications. "In ogni aspetto," wrote Emilio Morpugo, "la posta non rassomiglia più al passato, nelle forme e nel suo modo di operare, anzi nel suo carattere, essa è lo specchio del tempo, riflette in se stessa i suoi episodi più interessanti, ritrae con la memoria della sua vita la storia del popolo fra cui funziona e di cui è tanta parte" [In every aspect, the mail does not resemble what it was in the past, in its forms and operational modes, even in its character, it is the mirror of its time, it reflects its most salient episodes, it portrays with its history the history of the people for which it functions and of which it is a considerable component] (28). While Morpugo's words exemplify the rhetorical language – infused with Risorgimento pathos – used by nineteenth-century government functionaries, as well as by those who, after the unification of Italy, believed in the wondrous effects of progress, they nevertheless demonstrate an enthusiasm for letter writing as an instrument of cultural and social transformation, and as an opportunity for interaction until then afforded to only a few. Obviously, people did not begin writing letters in the nineteenth century, but it is only at this moment of the convergence of different economic, social, and cultural phenomena – with growing literacy rates, a wider newspaper circulation, and a more reliable postal service – that letter writing became an everyday practice. According to Jürgen Habermas, the traffic in news developed at the time alongside the traffic in commodities, and mail as a social practice can be said to have existed "only when the regular opportunity of letter dispatch became accessible to the general public, so there existed a press in the strict sense only once the regular supply of news became public, that is, again, accessible to the general public" (16). In Italy, this occurred in the second half of the nineteenth-century with the increase of social mobility and the emergence of a modern traffic in commodities and news. The middle of the nineteenth century represents a point of no return in the process of transformation of the means of communication.[24] And if sending and receiving letters had become by the second half of the nineteenth century a common aspiration if not a social necessity, there were also specific conventions – some old, some new – attached to epistolary writing with which correspondents were expected to comply in their private and public communication.

Figure 3: Back cover of *Calendario Postale 1898*. Archivio Storico di Poste Italiane, Rome, Italy.

Figure 3: (Continued)

Epistolary Manuals

By the second half of the nineteenth century, more and more people were writing letters and consulting epistolary manuals. The high number of books and book chapters on epistolary conduct published after Italy's unification attests to the existence of a growing number of consumers of this genre. If, as Roger Chartier suggests, letter writers were also often readers, then this "community of readers" became in the second half of the century a community of consumers ("Introduction" 2). Epistolary manuals became most popular at a time when conduct books and self-help literature generally were in vogue – that is, when society responded to social changes with cultural interventions that were supposed to stabilize and often contain such transformations. The production of conduct material typically intensifies during times of accelerated political and social change, and epistolary manuals, like most conduct books of the time, aimed at normalizing behaviour according to a new vocabulary of social relations. A relatively easier social mobility and wider access to education had prompted the production of a vast array of didactic material that revealed, in its obsession with detailing personal and collective behaviour, a sense of latent anxiety caused by what was perceived as too rapid – and possibly threatening – changes to the social order. As scholars of this literary genre have pointed out, conduct material naturally places its attention on the creation of harmonious social relations, but a closer look suggests that its main focus tends to be on those areas of social interaction that are filled with generational, gender, and class tensions (Turnaturi, *Signore e signori d'Italia* 13). In other words, if the main objective of conduct books is to produce social harmony, their narrative is imbued with overtones of conflict. Hence, it is a mistake to view conduct books – including epistolary manuals – as texts merely illustrating social etiquette, for behind their conventional surface of propriety they display a whole range of emotions connected to the fears and anxieties experienced by society at that particular historical moment.[25] Literary representations, whether drawn from the productions of high or low culture, constitute the outcome of social, historical, and cultural dynamics that are substantially conflictual. According to Franco Moretti, literature – and, I would add, popular literature in particular – "is perhaps the most omnivorous of social institutions, the most ductile in satisfying disparate social demands," for it has a fundamental function of securing "consent" (*Signs Taken for Wonders* 26–7). This was particularly evident in the nineteenth century when the print industry and cultural institutions more broadly took upon themselves the primary role of creating narratives of compromise between the different emerging social groups, and of harmonizing the differences by offering modalities of behaviour that would promote acceptance of dominant value and hegemonic rules (ibid. 27). This is not to say that readers of nineteenth-century popular fiction were merely passive recipients of cultural norms. On the contrary, a relatively wider social mobility and availability of cultural commodities created a readership comprising a varied typology of consumers' expectations. Maria Grazia Lolla's study of the politics and poetics of reading in nineteenth-century popular culture convincingly argues that the reader-writer relationship in fin-de-siècle Italy was transformed to the point that "the reader was feared as a most dangerous rival and held responsible for degrading literature and the literary experience" (22). No longer an external, acquiescent figure, the modern reader constituted in the eyes of many Italian writers a challenge, if not an obstacle, to their artistic autonomy. But publishers and authors could no longer neglect to take into account readers' expectations and consumer preferences when confronting their aspirations – whether of an artistic or a commercial nature – to develop a national literary tradition or cultural industry. Filled with conflict and anxiety, epistolary manuals as a subgenre of nineteenth-century conduct material constitute an eminent example of the kind of cultural interventions that sought to contain the sense of disruption perceived in society during the process of redefinition of social and cultural identities, and of transformation from an older order of life to modern social interactions. Because epistolary manuals strove to normalize behaviour by creating a shared language of sociability, they produced a blueprint of rules that responded to the willingness of the aspiring correspondents to learn but also enhanced their need to conform to a uniform standard of social propriety and cultural know-how. As an old adage goes, "Di necessità virtù" [Making a virtue out of necessity], or as Giovanni Mestica put it in 1874 in his *Istituzioni di letteratura*, the letter "servendo alle persone di ogni età e condizione, dall'umile artigiano all'uomo di stato, è il componimento più comune negli usi della vita e il più necessario" [serving people of all ages and conditions, from the humble artisan to the statesman, is the most common form of writing in life and the most necessary] (Mestica 127). The print media, through their many publications, expended much effort to turn their readers into "virtuous" practitioners of epistolary communication.

Epistolary manuals belong to what scholars have referred to as *istituzioni della socialità,* those social places and events "– la conversazione, il passeggio, le visite, la chiesa, la mensa, lo scrivere lettere – nei quali, nell'ordinario quotidiano, si ripeteva e riproponeva la rappresentazione di una personale presentabilità sociale" [– conversation, promenade, visit, church, lunch, correspondence – where, in the ordinariness of daily life, the demonstration of one's social respectability took place]; such institutions revolved around the court in the ancient regime and around the school and the family in modern times.[26] In the late nineteenth century, epistolary manuals were mainly read in school or at home and were published either as part of a textbook or as a commercial manual aimed at the whole family. They provided readers with a taxonomy of rituals and conventions that were supposed to improve people's writing skills and, at the same time, instil a sense of social order and cohesiveness, according to which correspondents would become not only individual practitioners of epistolary activity but members of a larger social group engaged in a modern network of communication. Seen in this light, letter manuals constituted a social "outing," a virtual space of connection between letter writers and the wider network of postal users. Moreover, they represented a cultural event that created knowledge and historical consciousness; from the solitude of his or her writing table, the reader of an epistolary manual was not only learning how to write a properly executed letter but was made aware of a literary tradition of epistles preserved throughout the centuries by the literary canon. A synthesis of literary references and practical suggestions, these manuals commonly offered a selection of exemplary letters, which were used for both their instructional quality and entertainment purposes. Whether drawn from the correspondence of famous literary figures or from contemporary epistolary exchanges, the letters included in these manuals offered an array of stories laden with inner and outer conflicts (from personal feelings of envy, solitude, or love pangs to more general problems relating to one's professional struggles and ambitions) that added a fictional aspect to their main function. The popularity of publications containing epistolary instructions – whether in the context of textbooks used in the classroom or of commercial volumes addressed to the whole family – cannot be understood apart from this double mandate, educational and entertaining, which shaped much of the cultural life produced in the aftermath of the country's political unification, when culture was understood to represent a main vehicle for the advancement of society. Enjoyed for their entertaining quality, epistolary manuals also engaged readers' historical consciousness by tracing a direct genealogy between past and present epistolary practices, thereby introducing the epistolary space of self-definition as a literary space of national representation.

The political unification of Italy represented a benchmark in the history of conduct books, and a similar historical shift may be identified in the publication of epistolary manuals before and after 1861.[27] Conduct books as well as epistolary manuals published in the second half of the century presented in terms of format and content an approach mindful of the changes taking place in the social and cultural profile of the readership and in the marketing and editorial policies adopted by the evolving print industry. Authors of the most popular conduct books emphasized the modernity of their texts, on the basis of a declared break from past conventions. As Marchesa Colombi put it in her *Gente per bene* (1877), identified by Inge Botteri as Italy's first "modern" conduct book,

> Cadono le città, cadono i regni, e cadono le costumanze adottate fra la gente civile. Ai tempi di Monsignor della Casa erano considerate inciviltà parecchie cose che ora sono ammesse. Invece non si troverà nulla nei galatei antichi sullo scambio delle carte da visita, sulle partecipazioni di matrimonio, nascite, morti, guarigioni [...] e tante altre cose che appartengono alle nostre usanze moderne. (4)
>
> [Cities fall, kingdoms fall, and customs used by civilized people fall. During Monsignor della Casa's times many things accepted today were then considered uncivilized. On the other hand, nothing will be found in those ancient conduct books on card exchanges, wedding announcements, birth, death, and convalescence ... and many other things that belong to our modern lifestyle.]

While both epistolary manuals and conduct material shared a common readership and strove to convey with their narratives a sense of modernity in the representation of new social mores and cultural identities, it would be a mistake, as the lack of studies focused exclusively on epistolary manuals implies, to think of epistolary manuals and conduct books as one and the same.[28] Authors of letter-writing manuals were, in fact, less willing than writers of more general conduct books to use, even only for rhetorical purposes, a narrative of clear differentiation between old and new practices of epistolary conduct. On the contrary, they made the connection with the past a central part of their books' appeal. If the circumstances and practices of modern communication had indeed changed by the late Ottocento, the epistolary expression for these writers remained essentially entrenched in an ideal cultural past, often in the literary canon. These writers' predilection for a historical approach to epistolary conduct material was not, however, devoid of ambiguity and contradiction. By promoting modalities of letter writing rooted in the literary tradition and often drawn from the correspondence of famous literary figures, these books melded current aspirations for modern patterns of communication with old paradigms of literary expression, projecting a sense of unity and continuity that was in sharp contrast not only to the general institutional commitment to modernize Italy but also to the heterogeneous reality (in terms of social, economic, and cultural composition) of the readership these authors were striving to influence.

Further clarification of this distinction between conduct books and epistolary manuals comes from two of the most popular conduct-book writers of the late nineteenth century, Marchesa Colombi and Emilia Nevers, who dismissed the importance of epistolary conventions as old-fashioned and therefore antithetical to the idea of modernity they advocated in their best-selling books. Marchesa Colombi devoted very little attention to epistolary conventions, and, in terms of form, praised simplicity over artifice: "Le lettere tengono luogo di discorsi. Scrivano come discorrerebbero e basta" [Letters take the place of speech. Write as you would speak and that's it] (82–3). A similarly dismissive approach towards epistolary rules may be found in *Galateo per la borghesia* (1883) – another best-selling conduct book with six editions in eight years – written by Emilia Nevers, who also opted for naturalness over stylistic formality in letter writing:

> Trovo anzi che nelle lettere – le quali rappresentano affatto la persona – ogni studio di rettorica, ogni ricercatezza nuoce, e invece di dar piacere, di far sorridere e di commuovere, fa esclamare a chi la riceve: Oh! Quanto deve aver sudato l'amico a mettere insieme tutta questa roba! [...] Che sia di stile schietto e spigliato, ecco quanto preme. (152)
>
> [I find rather that in writing letters – which represent the person as he or she is – any rhetorical application, any intentness is detrimental to the purpose of pleasing, making one smile, and moving the person who receives the letter and utters: "Oh, how difficult it must have been for my friend to put all this together!" ... What is important is that the style is frank and natural.]

Marchesa Colombi and Emilia Nevers addressed mainly a female audience – hence the notion that this genre went through a process of feminization in the second half of the nineteenth century. But though women did write letters, the practice itself was still considered to be essentially a male activity; for women to write letters outside of the strictly domestic sphere meant risking their reputation.[29] This may explain the reluctance of many women (as I discuss further in the next section of this chapter) to sign letters they sent to the editor of a newspaper, since exposing one's name to public view in a printed medium that circulated outside the family circle could be potentially damaging, not only to one's own reputation but to that of the whole family. In addition, since proper letter writing required a certain degree of formal education that was usually unavailable to women of the middle and lower classes, Marchesa Colombi and Emilia Nevers necessarily had to simplify the epistolary conventions for their intended female readers.

If Marchesa Colombi and Emilia Nevers minimized the relevance of epistolary conventions in their conduct literature, authors of epistolary manuals provided detailed rules that in principle reflected the scientific approach to pedagogy advocated by the positivist movement. Italian positivists and pedagogues, from Pasquale Villari to Roberto Ardigò and Aristide Gabelli, called for Italy's liberation from its backwardness through the creation of a more democratic distribution of knowledge among its citizens, and authors of epistolary manuals responded with a comprehensive compendium of epistolary rules as the necessary first steps for the implementation of a pedagogy aimed at "ringiovanire i costumi e la vita in modo consentaneo alla civiltà moderna" [renewing customs and life according to modern civilization] for a readership only recently emancipated from illiteracy (Gabelli, quoted in Spirito, ed., *Il pensiero*, 107). While epistolary manuals intended for school students presented in minute detail the theoretical and formalistic aspects of letter writing, the more commercial ones, like Cesare Causa's popular modern *secrétaire*, had a utilitarian approach and were filled with practical suggestions on topics ranging from the type of paper and handwriting to the address, salutation, and closing format to choose for every circumstance. Cesare Causa, *ex-garibaldino* and "scrittore emblematico del primo Salani" [a leading writer of the early Salani], had found in the publisher Adriano Salani a perfect partner for the creation of inexpensive books marketed for a public of newly literate readers eager to embrace culture as a means of social advancement (Faccioli 376).[30] Causa's commercial epistolary manuals were therefore conceived within the cultural trend and marketing efforts of self-help material.

The most prescriptive approach in matters of epistolary conduct may be found in school textbooks. Giuseppe Picci's *Guida allo studio delle belle lettere e al comporre con un manuale di stile epistolare* (1850) and Giovanni Mestica's *Istituzioni di letteratura* (1874–6) were manuals intended to be used by a public of high-school teachers and students. Both Picci and Mestica presented letter writing as an integral part of the study of rhetoric, in particular of the *art dictaminis,* and they offered a modality of epistolary writing that was deeply rooted in the classical literary tradition. No doubt a sense of utilitarian purpose informed these textbooks as well, but their emphasis was more on the formal rather than the practical dimension of letter writing. The rhetorical aspect of the epistolary tradition became the main organizing principle of the pedagogic mission of these texts aimed at reforming the school system and at reawakening the historical and civil consciousness of society (Bertoni Jovine, *Storia dell'educazione popolare* 21).

Picci's *Guida alla studio delle belle lettere e al comporre con un manuale dello stile epistolare* [Guide to the study of letters and composition with a manual on epistolary style], written in 1850 and in its eleventh edition by 1901, was printed simultaneously by Oliva in Milan and Paravia in Turin, both publishers who specialized in textbooks. While this manual was originally written before the political unification of Italy, the height of its popularity – evidenced by the many editions published in the course of half a century – was reached after 1861. Divided into three parts – the first devoted to writing in general, the second to prose, and the third to poetry – Picci's guide to composition provided simple but effective instructions on writing. The section devoted to epistolary writing, included in the second part of the book, offered a comprehensive summary of epistolary conventions indexed according to a list of rules and occasions, from letter closings (*chiuse di rispetto* or *chiuse di amicizia*) in formal and informal correspondence, to personal titles to be used with different correspondents (*titoli pei secolari* or *titoli per gli ecclesiastici*), to specific styles and wordings used to express sympathy or congratulations or to request a favour or advice from someone.

Picci's epistolary conventions, though, were not merely practical rules. Infused with historical significance, epistolary writing marked for Picci the very origin of the Western literary tradition. "Il genere epistolare," Picci wrote, "siccome creato dalla necessità di comunicare a' lontani i nostri pensieri, deve naturalmente essere antichissimo, fin nella Bibbia e nell'Iliade n'è fatta menzione" [The epistolary genre, given that it was created by the need to communicate our thoughts to those living far from us, must naturally be very ancient; it is already mentioned in the Bible and the Iliad] (299). This historical genealogy, while naturally serving the primary purpose of bestowing prestige upon the epistolary genre, rhetorically delivered a message of national pride and, ultimately, civic responsibility on the basis of a presumed common cultural heritage traceable to ancient Roman times. The brief listing of notable letter writers included by Picci (Cicero, Pliny the Younger, Petrarch, Bembo, Giovanni della Casa, and Leopardi, among others) pointed directly to this historical and cultural patrimony, in addition to calling the reader's attention to the close correlation that had always existed between individuals and their epoch, between people's civil conduct and the glory of the era in which they lived: "[D]a ciò la loro importanza come sincero testimonio degli alti pensieri e degli onesti e teneri affetti dell'autore, come documento verace della sua vita pubblica e privata e della storia de' suoi tempi" [From this comes the sincere testimony of the lofty thoughts and honest and tender affections of the author, as a true document of his public and private life and of the history of his time] (299). From this brief but essential historical preamble, Picci expanded on a detailed description of the rules regulating letter writing, organized thematically according to subdivisions related to specific situations ("Lettere di condoglianza e conforto," "Lettere di congratulazione e lode," "Lettere di consiglio," "Lettere d'invito" [Letters of sympathy and support, Letters of congratulations and praise, Letters of advice, Letters of invitation]). At the end of each subdivision, a sample of letters selected from the literary tradition – letters by Cicero, Gozzi, Tasso, and Leopardi, among others – were presented to elucidate the stylistic and rhetorical points made by the author. Among the letters chosen to illustrate the "lettera di rimprovero" [letter of reproach], for instance, Picci included a letter by Pietro Bembo to his nephew Carlo, who, dodging the responsibilities usually expected for his age, was failing to become the kind of youth ("valoroso e virtuoso e dotto" [brave, virtuous, and learned] [545]) that his family wanted him to be. Introduced by Picci as being exemplary, Bembo's letter was praised in particular for its moderate tone and harmonious style ("grave senza essere declamatorio, austero senza acerbità, mostra il linguaggio della ragione non della passione" [solemn but not declamatory, stern but not acerbic, it shows the language of reason and not of passion]), and for its ability to deliver advice without the harsh tones of personal conflict – an ideal of moderation (*medietas*) that Picci adopted from the Renaissance and applied to his own time to promote restraint and harmony in social relations (546). Classical letter writing was, in this sense, presented by Picci, as an instrument still effective in modern times in shaping not only personal correspondence but also social relations. Finally, a bibliography referencing epistolary manuals, including *L'arte di scrivere lettere dedotta dall'analisi dei classici per opera di Giuseppe Ignazio Montanari* (1840) and Gasparo Gozzi's *Segretario moderno, o ammaestramenti ed esempi per ogni sorta di lettere tratti dai più illustri scrittori* (1820), concludes the chapter on epistolarity as a reminder that letter writing as a social activity ought to be conducted according to the rules established by the literary tradition and valued for its positive moral effects on contemporary society.

Another manual, similar in tone and format to Picci's, is Giovanni Mestica's *Istituzioni di letteratura*, published in Florence by Barbera between 1874 and 1876 in two volumes, with a chapter in the second volume of 129 pages devoted exclusively to epistolary writing. Dubbed by critics the "canto del cigno della vecchia tradizione retorica" [swan song of the old rhetorical tradition] (Romagnoli 279), Mestica's book followed the principle that "poiché la lettera tiene luogo della conversazione fra persone civili, deve anche obbedire agli usi e alle cerimonie stesse che nel parlare insieme osserviamo" [because the letter takes the place of a conversation among civilized people, it must also respond to the use and ceremonies observed when we talk to each other], although when talking about using epistolary conventions, the author referred principally to the modalities of letter writing inherited from the past rather than those shaped by more current social practices (II, 4). For Mestica, epistolary conventions belonged essentially to the literary tradition and gained social value from the prestige bestowed on them by famous writers: "I modelli di scriver lettere ci sono porti dagli epistolari d'insigni prosatori e poeti [...] Se non che gli epistolari di qualche mole appartengono tutti ai grandi scrittori, ed hanno pregio non solo come monumenti del bel dire, ma eziando perché servono d'illustrazione alla vita e alle opere degli autori stessi ed ai tempi nostri" [Modalities of letter writing are given to us by the correspondence of great novelists and poets ... Letters of note belong to great writers, and have value not only as monuments of eloquence but also to illustrate the life and works of those authors in our times] (II, 12–13). Classical modalities of epistolary interaction remain valid on the basis of the assumed universality of the modes of expression used by literary figures, whose epistolary exemplarity stems essentially from the authority of their literary achievements. Historical letters retain semantic value insofar as they belong to a larger corpus of literary production, thus highlighting the public cultural import of letter writing. As Mestica eagerly pointed out: "Gli autori, le cui lettere ci servono di modelli, sono tutti insigni per altre opere in prosa e poesia per lo più di assai maggiore importanza e di grande perfezione" [The authors whose letters we use as models are all illustrious for their other works in prose and poetry of greater importance and perfection] (II, 16).

Mestica's compendium of epistolary norms was conceived mainly as a manual of rhetoric to be used in the classroom for the "scuole mezzane," an expression by which Mestica referred to the *ginnasi, scuole normali,* and technical secondary schools (I, iv). What distinguished his rhetorical treatise from others, to the point that it was noted by Giosuè Carducci in his 1880 report to the Ministry of Education as one of the few recommended textbooks of rhetoric (Raicich 153), was the scope of the rhetorical discourse engaged by Mestica, who saw in this educational enterprise an opportunity to improve not only students' competence in writing but also their overall academic performance as students in a newly reformed school system. One of the most urgent tasks of the newly constituted Italian government was to create a sense of national consciousness – a task primarily assigned to the recently created national school system. Even though the lack of sufficient funds and the dearth of well-trained teachers soon revealed the limits of a vision of the school as a panacea for all Italian ills, school programs and textbooks relied heavily on such a rhetoric of cultural reform.

Mestica's *Istituzioni di letteratura* made the rhetorical discourse thematically and structurally central to its epistolary narrative. For Mestica, rhetoric – whether oral or written – is essentially dialogic in nature, meaning that the presence of an implicit or explicit interlocutor who is to be persuaded functions as an integral element of the performative act: "l'arte del dire a fine di persuadere, o, nello stesso ristretto senso, semplicemente arte del dire [...] E oggidì si suole anche più generalmente chiamare arte di scrivere" [the art of speaking with the objective of persuading someone, or, in the strict sense, simply the art of speaking ... is today more generally called the art of writing] (I, 7). Hermeneutically constructed on a similar cause-effect dynamic, Mestica's epistolary discourse privileges the dialogic rather than monologic or expository form of expression: "poiché noi scrivendola conversiamo realmente con la persona, a cui vogliamo indirizzarla" [as when we write it we really converse with the person, to whom we are addressing it] (II, 3). To write a letter, Mestica suggested, required both inward and outward observation, good self-expression accompanied by the understanding of one's interlocutor's nature. For Mestica, finally, epistolary writing was a "conversazione tra assenti" [conversation between absentees] (II, 2), an expression inherited from the Ciceronian tradition, *amicorum colloquia absentium*. Recent rhetorical studies have paid close attention to this aspect, pointing out that "Rhetoric is a not a manner of performing, but of 'observing' what are, in every situation, the available means of persuasion" (Valesio 28). For Mestica, too, persuasion is central to the rhetorical discourse:

> La lettera, dunque, è un componimento, col quale si fa intendere a chi è lontano ciò che gli avremmo detto a voce se fosse stato presente [...] Nondimeno considerando che i colloqui, che facciamo, riguardano per lo più le azioni, e mirano a persuadere o a dissuadere, da ciò segue che la lettera debba principalmente rapportarsi al genere persuasivo e in particolar modo a quella che dicesi eloquenza per gli usi del viver comune in privato e pubblico. (II, 2)
>
> [The letter, therefore, is a composition with which we communicate to those who are far away what we would have said if they were present [...] Nevertheless, considering that our conversations refer mainly to actions, and aim at persuading or dissuading, it can be inferred that the letter has to be principally related to the persuasive genre, in particular to what is called eloquence, for the uses of common life, whether private or public.]

The persuasive aspect of letter writing invested, as Mestica pointed out, both the private and public realms of life; it was particularly relevant, as will be discussed later in this chapter, to the public discourse (for which this type of manual was intended as a training instrument) constructed on the basis of specific epistolary narratives and conventions aimed at influencing one's interlocutors and, ultimately, at creating public opinion. Students of the *ginnasi* or *scuole normali*, being trained to become the future leaders of the nation, were thus taught how to communicate their ideas effectively in a political, educational, or cultural environment.

Mestica's treatise on epistolary writing is divided into twenty-four sections, each one combining erudite illustration of stylistic norms with quotations from literary sources. In Section 7, for instance, "Delle lettere familiari attinenti al genere persuasivo; lettere officiose" [Familiar letters relative to the persuasive genre; letters for special occasions], Mestica introduced the concept of persuasiveness, noting that this type of letter is the most difficult to write. Given the natural human tendency to be overtly sentimental, pompous, or adulatory when writing letters, Mestica warned, it was necessary for the reader to firmly rely on epistolary exemplars from the past in order to avoid the potential pitfalls of sentiment and emotionalism. In a tenuous balance between naturalness and artifice, traditional and modern demeanour (*urbanità*), Mestica's epistolary manual interwove new and old epistolary narratives, providing a commentary on proper epistolary communication interspersed with excerpts drawn from letters by Cicero, Machiavelli, Galilei, Leopardi, and Francesco Redi, among others. Such alternation of historical texts and instructional material produced what Foucault defined as an *archive* of historical statements, an epistolary discourse characterized by breaks and discontinuities rather than by a coherent organic unity that created a social rubric of, in Mestica's case, epistolary practices (*Archeology of Knowledge* 130–1). For Mestica, the epistles included in his anthologized epistolary treatise were not relevant insofar as they were representative of the work produced by the authors who wrote them, but rather constituted literary units that gained semantic value only once they became part of a metanarrative of epistolary Italian history that was to be learned in modern times.

> Per tale rispetto, gli epistolari non solo servono d'illustrazione e di compimento alla storia, sia nel campo de' pensieri, sia in quello delle azioni, ma vengono a formare una storia tutta propria di queste e di quelli, la quale tanto più è fedele rappresentazione del vero, quanto più va nel concreto e in quelle particolarità e minuzie, donde risulta il carattere della vita intima ed esteriore degli individui. Quale istoria, per esempio, ci potrebbe guidare con tanta sicurezza in mezzo agli avvolgimenti e alle condizioni reali dei tempi di Cicerone? (II, 128)
>
> [In this respect, letters serve not only the purpose of illustrating and completing history, both in terms of creation of thoughts and actions, but themselves create a history which is most faithful to the representation of reality, insofar as it delves into the concrete part of life with all its particularities and minutiae, from which the character of the intimate and exterior life of the individual unfolds.]

Because letters had the capacity to chronicle history as it unfolded, Mestica used them as an instrument with which to gauge the reality of his relationship with his own times, and more particularly with his readers. If Cicero's letters not only represented the great Roman times but also made history themselves, what kind of historical account were these same letters producing when read within the context of late-nineteenth-century reality? Letters, as Janet G. Altman points out, are inherently apt to produce multiple signifying variations: a single letter can be addressed to one or more recipients, and that same letter can later be readdressed within a corpus of a correspondence or in an anthology to multiple audiences. For Altman, letters are both "autobiographically undressed" and "rhetorically addressed"; they are both private discourse and public speech, and when published in a volume, they can be "readdressed to a new readership" and "redressed" or corrected by the publisher ("Letter Book" 19). In other words, epistolary texts lend themselves easily to the construction of multiple narratives or of a metanarrative aimed at historicizing the very process of reading and writing. For both Mestica and Picci, the historicity produced by their selection of literary epistles addressed and redressed the educational concerns formulated by pedagogues and institutional figures who identified in history itself – in its various forms of representation – the main educational instrument in the process of reformation of Italy's backward school programs. In light of this historical approach, therefore, Cicero's epistles created historical memory and Mestica and Picci used these epistolary documents to provide their students with a lesson about historical engagement and, ultimately, civic responsibility, as not only pertinent but crucial to the reality of contemporary life.

Regardless of how unconcerned these authors appeared to be with the practical realities of modern communication, their epistolary manuals, infused with deep historical implications, followed the principle, generally ascribed to Francesco De Sanctis's pedagogical theories, that school represented the main moral medium and cultural forum for the improvement of the individual and, by extension, the whole society, for it facilitated not only the acquisition of factual knowledge and professional training but, most importantly, the development of personal character; that is, "non vale solo a educare l'intelligenza, ma, ciò che è più, ti forma la volontà" [it is effective not only for educating the intelligence, but also, and most importantly, for forming one's will] (De Sanctis 280). By choosing history as the privileged site in which the modern epistolary subject ought to emerge, Mestica and Picci reaffirmed the fundamentally conservative nature of their educational project, in which the past, with all its pre-established parameters, remained incontestably the main template for the definition of a modern sense of national consciousness. Ultimately for Mestica, as for Picci, epistolary conventions were rooted in tradition, frozen in an indefinite past and yet adaptable to changing times.

By the end of the nineteenth century, epistolary manuals were also commonly featured in publishers' commercial catalogues. As Giuseppe Fumagalli put it, in his introduction to Jacopo Gelli's *Come devo scrivere le mie lettere? Esempi di lettere e di scritture private per tutte le occasioni della vita* [How to write letters? Models of letters and private writing for all life occasions], published in 1898 and reprinted several times until the 1930s,

> Purtroppo la lettera tiene oggi nella vita umana un posto considerevole, assai più che in quella dei padri nostri, senza risalire alle generazioni anteriori. E le cause sono molte e complesse. Si viaggia e ci si muove oggi assai più facilmente che cinquant'anni fa, sia per affari, sia per diporto; poiché grazie ai progressi della scienza il viaggiare è più comodo e costa meno, quindi le famiglie, le relazioni si sparpagliano come semi abbandonati al vento [...]. Inoltre oggi il convezionalismo, la moda, la diffidenza, vogliono che molte faccende, anche di poco momento che i nostri nonni sbrigavano a voce, in quattro parole, si trattino più prudentemente per lettera. Vuoi invitare una persona di riguardo a pranzo? Scrivi. Vuoi chiedere qualcosa a un pubblico ufficio? Scrivi. Vuoi fare una dichiarazione di stima, o diffidare alcuno, comporre un dissidio? Scrivi. (x)
>
> [Unfortunately letters play a considerable role in today's life, much more than they did for our fathers, to say nothing of previous generations. And the reasons are multiple and complex. We travel and move around much more easily than fifty years ago, for both business and leisure, because thanks to the progress of science travel is more comfortable and it costs less, hence families, relationships, spread around like seeds dispersed by the wind. Moreover, customs, fashion, caution, all require that many of our actions, even those of little account that our grandfathers resolved quickly orally, are treated more prudently by letter. Do you want to invite a distinguished person for lunch? Write. Do you want to ask something of a public officer? Write. Do you want to make a declaration of esteem, or give notice to someone, or resolve a conflict? Write.]

Writing letters had become a common practice, and publishers saw in this genre a source of commercial profit. Publishers' catalogues were filled with long, wordy titles of books devoted to the art of writing letters, manuals that were seen as having "la massima utilità per tutti" [mass utility] – "per tutti," here, signifying the middle and the lower classes, previously excluded by conduct books such as Melchiorre Gioia's *Nuovo Galateo* (1820), which, though infused with reforming ideals inspired by the Enlightenment, was still addressing mainly an elite readership.[31] The popular orientation of these new publications is revealed by the abundance of information – providing minute details, such as what paper to use, how to address people, and so on – offered to a public of novice readers and writers unfamiliar with the rules of good epistolary conduct. As Edmondo De Amicis somewhat scornfully described in his short story "Il libraio del ragazzo" [The bookstore for youth], commercial books with little literary merit sometimes filled the windows of bookstores: "Compiva la bizzarria di quella mostra una piccola flora libraria destinata alle serve che accompagnano alla scuola i ragazzi: una serie di volumetti zozzi e plebei, dalle copertine verbose come *La vera chiave del Tesoro, La cuoca piemontese e il Segretario galante*" [A small group of books targeting maids who accompanied children to school completed that bizarre window exposition: a series of dirty and plebeian volumes, with verbose titles such as "The True Treasure Key," "The Piedmontese Cook," and "The Gallant Secretary"] (De Amicis 4).[32] This latter book, *Il segretario galante ossia modo di scrivere lettere amorose sopra ogni sorta di argomento* [The gallant secretary or how to write amorous letters on any sort of topic], by Cesare Causa, was not only published in several editions between the 1880s and the early twentieth century but became the prototype for many other books, which, published under slightly different titles and not uncommonly omitting the author's name, reproduced basically the same content, with minimal variations. A *Segretario galante italiano-inglese* was even published by Salani in 1913, presumably for the English-speaking community living in Florence and, according to Mary Anne Trasciatti, Causa's *Segretario galante* was enormously influential in the composition of *Manuale di corrispondenza familiare, commerciale, amorosa italiano-inglese*, written in 1909 by Alfonso Arbib-Costa, a professor in Romance languages at the City College of New York, "to provide models of effective written communication for [Italian] immigrants looking for social acceptance" and to facilitate "the formation of romantic relationships across ethnic and linguistic boundaries" (74–5).

Piccolo segretario galante o raccolta di lettere e biglietti amorosi aggiuntavi la corrispondenza di due infelici amanti (1882)

Il vero segretario galante contenente il metodo di scrivere lettere amorose sopra diversi argomenti aggiuntovi l'epistolario di amanti celebri ed altri scritti galanti (1887)
Il segretario degli amanti per imparare a scrivere. Lettere e biglietti amorosi, nonché lettere di discordia e di accomodamento, aggiuntovi la scuola d'amore e molti segreti infallibili per gli innamorati (1895)
Modernissimo segretario galante: Raccolta di lettere (1921)
Il moderno segretario galante (1926)
Il moderno segretario galante o l'arte di farsi amare (1935)
Il moderno segretario galante: Lettere d'amore (1937)
Il moderno segretario galante con l'aggiunta di lettere ai militari (1940)

These were all manuals modelled after Causa's original *Segretario galante* and reproducing long lists of possible epistolary exchanges among lovers, with detailed descriptions of formal aspects of letter writing (paper, style, conventions) as well as of circumstances that would require an exchange of correspondence.[33] In some cases they included also models of letters taken from the literary tradition (*Abelard and Eloise, Jacopo Ortis, Nouvelle Eloise,* and so on) or reflecting the social conventions pertaining to courting (i.e., a letter proposing marriage) and amorous discourse in general. Causa's commercial success with this genre, however, may not be the result solely of the popularity of the epistolary manual as a genre. His fame owes much to his ability to interpret the public's current reading tastes, cultural expectations, and learning ambitions, for he was also the author of a popular series of comedies centred on the comic Stenterello character, and of several biographies of historical figures of the Risorgimento, such as Victor Emanuel II, the Bandiera brothers, and Ciro Menotti.[34] Indeed, looking at the Salani catalogue, one can see that, though Adriano Salani was already selling *segretari* in the early 1860s, it was only when Causa began writing in this genre that it became a major commercial success for the Florentine publisher.[35] Given the number of editions published in the latter part of the nineteenth century and the ubiquitous presence of this manual in bookstore windows, one can only conclude that Causa's cultural know-how and Salani's commercial acumen were a timely response to readers' growing need to engage in correspondence and their growing belief in the power of cultural instruction to help them improve their social status.

Commercial epistolary manuals commonly featured the love letter and included elements of fictionality that were supposed to make the reading more entertaining. A fictional aspect of *Segretario galante*, for instance, may be found in a chapter titled "Corrispondenza per giornali" in which readers were offered instructions on how to conduct a clandestine amorous correspondence with the help of newspaper ads.

> Un nuovo mezzo ingegnoso, quanto bizzarro, si è da poco introdotto in Italia, quello cioè di scriversi mercè gli *Annunzi* di quarta pagina di un giornale onde mantenere, con chi non vuol far pervenire direttamente i propri caratteri, una corrispondenza amorosa clandestina, che non può che ritenersi in parte come scandalosa, non pertanto può servire anche ad onesti fini. (191)
>
> [A new, ingenious, but rather bizarre, medium has recently been introduced in Italy in order to correspond with someone through the ads on the fourth page of newspapers. A medium to keep a clandestine amorous correspondence with someone to whom one cannot write directly, which can be regarded as only partly scandalous, but which nevertheless can also serve honest aims.]

Causa explained that with a relatively small sum of money, from fifty cents to one lira, a person could have printed an anonymous newspaper message that his or her lover could read without raising any suspicion among friends or family members. "Vi sono amori proibiti mantenuti in tal guisa vivi e per lungo tempo senza che alcuno sia mai riuscito a trapelarne il segreto" [There are forbidden loves that are kept alive for a long time without anyone discovering the secret] (191). Similar instructions were offered by Jacopo Gelli, who concluded his *Come devo scrivere le mie lettere?* with a chapter titled "Corrispondenza segreta," which was filled with encrypted messages to be used in newspapers (see figures 4 and 5).While it is hard to determine how widespread this practice was, given the secret nature of the amorous enterprise, the simple fact that many of these manuals mentioned it demonstrates how this type of popular literature capitalized on sentimental narratives, and how it titillated the collective imagination and the public interest in transgressive sentimental plots, as the picture next to the title of Causa's *Segretario galante* clearly suggests (a man furtively delivers personal correspondence to a woman over the wall of a private residence; see figure 6). "Secrecy versus the family network," as Roger Chartier explains: "For a time, the conventions and guidelines of secretaries are forgotten, so that passion, sorrow or fantasy may be given an individual

Figure 4: Title page of Jacopo Gelli's *Come devo scrivere le mie lettere?* Milan: Ulrico Hoepli, 1900. Biblioteca Nazionale Centrale di Firenze, Italy.

146 *Come devo scrivere le mie lettere?*

Parole	Cifre	TESTO (significato)
Fugace . .	23	Rispondetemi per telegrafo alla mia lettera odierna.
Subito . .	24	Impossibile aderire alle vostre richieste per le ragioni che vi comunicherò con lettera o a voce.
Domani .	25	Oggi.
Ottobre . .	26	Domani.
Maggio .	27	Ieri. ESEMPIO: *maggio*; oppure: 27.
Sera . . .	28	Domani l'altro.
Appetito .	29	Ieri l'altro.
Spavento .	30	Ogni ricerca è riuscita infruttuosa; comunicatemi nuove istruzioni.
Cotonificio	31	Sarò costì alle ore.... trovatevi in stazione. ESEMPIO: *Cotonificio*, 19; oppure: 31 19.
Stampa .	32	Partirò.... (*giorno*). ESEMPIO: *stampa venerdì*; oppure: 32 *venerdì*.
Nato . . .	33	Rimessa partenza a.... (*giorno*) alle ore....
Testa . .	34	Siete minacciato; non siete sicuro, ponetevi in salvo.
Speranza .	35	Vi attendo a.... (*nome della località*) Hôtel.... (*nome dell'albergo*) alle ore....
Fallacio .	36	Non vi fidate, prendete le vostre precauzioni e disponetevi a porvi in sicuro.
Festa . .	37	Mi trovo in una posizione tristissima.
Lutto . .	38	Tutto per il meglio; mi trovo in ottima posizione.

Corrispondenza segreta o Crittografia 147

Parole	Cifre	TESTO (significato)
Cordiale .	39	Verrò a trovarti.... (*giorno*) alle ore....
Spirato. .	40	Vi ho scritto fermo in posta a....
Lampo . .	41	Vi ho spedito oggi lire....
Giallo . .	42	Sto male....
Vivace . .	43	È morto oggi (*seguito da ?*: significherà ieri).
Azzurro .	44	È morto stanotte (*se con ?*: iernotte).
Riposo . .	45	Oggi hanno trovato morto il signor.... (*nome*).
Riparo . .	46	Si è suicidato.... (*nome*). ESEMPIO: *Giovanni riparo*: oppure: *Giovanni* 46.
Rosso . .	47	Debbo andare da (o a).... (*nome*) cosa devo dire?
Amaranto	48	Che cosa devo fare?
Luce . . .	49	Consigliatemi.
Saluto . .	50	Parto felicissimo. Bambino (*se con ?*: bambina) e madre stanno benone.
Treno . .	51	Mandatemi le vostre istruzioni a....
Verde . .	52	Vi attendo a casa mia oggi (*se con ?*: domani) alle ore....
Bianco . .	53	Annunziate la mia visita a.... (*nome*) per domani alle ore....
Carta . .	54	Vi attendo a.... oggi alle ore....
Romanza .	55	È inutile persistere....
Giovane .	56	A qualunque costo combinate.
Figlio . .	57	Domandate formalmente se sarei accolto per marito (*con ?*: moglie).

Figure 5: Jacopo Gelli's *Come devo scrivere le mie lettere?* Milan: Ulrico Hoepli, 1900, 146–7. Biblioteca Nazionale Centrale di Firenze, Italy.

IL

SEGRETARIO GALANTE

OSSIA

MODO DI SCRIVERE LETTERE AMOROSE

SOPRA OGNI SORTA DI ARGOMENTO

AGGIUNTOVI

L'EPISTOLARIO DEGLI AMANTI CELEBRI

ED ALTRI SCRITTI GALANTI

PER CURA DI

CESARE CAUSA

FIRENZE

ADRIANO SALANI, EDITORE

Via S. Niccolò, 102

1887

Figure 6: Title page of Cesare Causa's *Il segretario galante ossia modo di scrivere lettere amorose*. Florence: Salani, 1887. Biblioteca Nazionale Centrale di Firenze, Italy.

and subjective voice" – secrecy, that is, as a theme that permeated contemporary cultural discourse based on family affairs, as the vast corpus of nineteenth-century operas and domestic novels amply demonstrated ("Introduction" 21). As for much of the popular literature of the time, entertainment was seen as an effective tool for delivering a prescriptive message; it was not uncommon for these books to be read aloud to the whole family, and certainly the fictional elements woven into the formulaic descriptions of these manuals facilitated the performative aspect of reading for the enjoyment and instruction of a group of listeners.

Also popular was Cesare Causa's *Il segretario italiano ossia modo di scrivere lettere sopra ogni sorta di argomenti per affari di commercio, amici e conoscenti, per feste e anniversari di famiglia, lettere di raccomandazione, di domanda ecc. con l'aggiunta di modelli per cambiali, suppliche e ricevute, ricorsi al re, ai ministri ed altri funzionari e la spiegazione grammaticale per scrivere ben corretto e con buona ortografia* [The Italian secretary or how to write letters on all sorts of subjects for commercial affairs, friends and acquaintances, holidays and family anniversaries, letters of recommendation, application letters, etc. with additional models for promissory notes, requests and receipts, appeals to the king, ministers and other functionaries, and with grammatical rules for writing well and in the correct spelling] (1883), which had several anonymous reprints. These books represented the latest trend in cultural consumerism: the publication of volumes intended to be used for practical purposes by the whole family. Presented by the publisher/editor according to the latest marketing strategies, these books were sold not on the basis of the author's reputation but of the promise, announced in the title, of helping the reader achieve his or her social ambitions. In other words, the credibility of these books simply depended on the word "segretario," which inevitably appeared in all the anonymous reprints of Cesare Causa's prototype, often simply titled "*Segretario italiano*" or "*Segretario galante.*" These manuals usually addressed all members of the household, with the "*Segretario italiano*" providing a chapter on business letters for the male readership and a chapter on personal letters for female readers. Similarly, the "*Segretario galante*" inferred the presence of a mixed readership by its inclusion of both men and women as senders and receivers of love letters. However, the 1903 edition of this manual, published by Salani without the author's name, includes an image next to the title page of a man, sitting at his desk and taking care of correspondence, indicating visually that the primary intended recipient of the manual was the male head of the household (see figure 7). Women could write

IL

SEGRETARIO ITALIANO

OSSIA

MODO DI SCRIVER LETTERE

sopra ogni sorta di argomenti

PER AFFARI DI COMMERCIO

AMICI E CONOSCENTI

PER FESTE ED ANNIVERSARI DI FAMIGLIA

LETTERE DI RACCOMANDAZIONE, DI DOMANDA EC.

Con l' aggiunta: di Modelli per Cambiali, Suppliche e Ricevute, Ricorsi al Re, ai Ministri ed altri funzionari; e la spiegazione grammaticale per scrivere ben corretto e con buona ortografia.

FIRENZE

ADRIANO SALANI, EDITORE

Viale Militare.

Figure 7: Title page of *Il segretario italiano ossia modo di scriver lettere sopra ogni sorta di argomenti.* Florence: Adriano Salani, 1903. Biblioteca Nazionale Centrale di Firenze, Italy.

letters, but only under the careful supervision of male members of the family. The description of the content – a litany of practical advice about letter writing (including information about postal tariffs), grammatical and stylistic rules, as well as model letters for personal and commercial correspondence – reveals the self-help nature of this type of publication, conceived to give families practical guidelines but also to enable readers to forge new individual and collective identities as practitioners of modern cultural and social activities.

Whether included in textbooks for high-school students or published in volumes marketed to a mass readership, epistolary manuals of the late nineteenth century are indicative of a growing interest in the niceties of a social practice that was thought of as increasingly relevant to the conduct of private and public affairs. Lower postal tariffs, improved educational opportunities, and increased social mobility all helped to enhance the cultural and social significance of letter writing. Fumagalli highlighted the contrast with earlier days: "Tutto questo parrebbe un sogno ai nostri buoni nonni per i quali ricevere una lettera era un avvenimento non sempre gradito" [All this would seem a dream to our grandfathers, for whom receiving a letter was not always a pleasant event] (Gelli x), referring to the material difficulties people used to encounter when writing, sending, or receiving a letter. When the nationalization of the postal service paved the way for more efficient and affordable postal communication, people began to correspond more frequently and consulted epistolary manuals to be guided in the art of writing personal and professional correspondence. The increase in the volume and range of letter writing was seen not only in the realm of personal and commercial correspondence; the letter format was also sometimes a medium for public exchanges of information in the burgeoning newspaper industry of post-unification Italy.

The Print Media

Letters were a regular feature in the print media of nineteenth-century Italy, especially in periodicals devoted to a female readership. Imported from France towards the end of the eighteenth century, the practice of publishing letters in newspapers became more common in the next century. The reasons for the increase are many and are generally connected to the transformation of the print media and, more specifically, the transformation of the relationship between their production and consumption as cultural products, as outlined by Robert Darnton in his analysis of what he described as "a communication circuit that runs from the author to the publisher [...] the printer, the shipper, the bookseller and the reader" (11). The possible interconnected reasons include: the enhanced educational mandate of journalism (inherited from the Risorgimento period); the slow but increasing democratization of cultural industries, resulting in the publication of a wider variety of cultural products and in the growth of a public of readers and consumers; and increased competition among publishers and newspaper writers for an ever-larger share of the growing (though still rather small) market for cultural products.[36] From a strategic point of view, letters facilitated the delivery of a didactic message (addressed to an ideal receiving "you"), fostered a conversational approach to communication (by adopting a supposedly natural and spontaneous language), emphasized the role of the reader (by way of what Altman defines as "the epistolary pact – the call for response from a specific reader within the correspondent's world" [*Epistolarity* 89]), and evoked the presence of a community of readers composed of an idealized Italian readership (still in formation) and of real readers who, though dispersed across the newly politically unified nation, found here an opportunity to interact within a network of information exchange. Far from being merely ornamental, the formal aspect of these articles significantly influenced the production of meaning insofar as both readers and writers recognized in the letter a viable instrument that facilitated the creation of a direct, seemingly unfiltered, communication between them (ibid. 4). Writers not infrequently published articles on current issues in the form of letters addressed either to the director of the newspaper or to other known public figures. But it was not only writers who used the epistolary form for their communication; readers did as well. Newspaper editors and writers (often one and the same person) encouraged their readers to write to them, for through their letters they could gauge who their readers were, including their consumer tastes, opinions, expectations, and, implicitly, attitudes towards the paper. This is not to say that readers were merely passive recipients of editors' marketing strategies; rather, readers, perhaps for the first time, were given the opportunity to be part of the journalistic equation, and editors had to take their opinions into account if they wanted their newspapers to survive. As Giuseppe Bianchetti pointed out in 1858, "Per far vivere un periodico non tanto ci è bisogno di chi lo scrive quanto di chi lo legge" [For a periodical to survive, it needs more readers than writers] (84). And the letters published in newspapers constituted the very embodiment of an audience, the written evidence that a body of readers truly existed. In post-unification Italy, when the circulation of information at the national level increased, the letter cogently exemplified, with its innate property of overcoming distances while fostering communication, the natural aspiration of a burgeoning Italian journalism to reach a national audience that was perceived as physically and culturally distant and still fragmented.

This section will focus on two main practices of newspaper letter writing: letters to the editor written by regular citizens, included in columns sometimes called *Piccola posta, Conversazioni,* or *Corrispondenze;* and letter-articles – often referred to as *Lettera aperta* – published by well-known public figures, who wrote about current issues of interest to women, such as education, school reform, changing social identities, and gender roles. The purpose of this analysis is twofold: first of all, to identify a public of female readers (the same public who presumably read the fiction analysed in the following two chapters) and letter-writers, who gained access, through correspondence, to the public cultural sphere of late-nineteenth-century Italy; and second, to shed some light on both the content and form of this epistolary activity, which involved women living in the public spotlight as well as those inhabiting the larger, usually anonymous, domestic domain. The preceding section of this chapter suggested that institutional and commercial instructions on how to write letters were mainly addressed to a male audience – whether of students or heads of families. However, the research on letters published in periodicals devoted to a female readership shows that in them women found a space for information exchange and opportunities to be part of a national cultural network of readers and writers by means of a gender-specific discourse focused on producing public opinion among women.

As a virtual space of female sociability, letter writing in newspapers is historically connected to salon life, a cultural institution also revolving around the female figure.[37] The nineteenth-century salon or, more precisely, the "salotto di conversazione italiano," though not comparable to that of France, also flourished for a time in Italy, and shared several common elements with the print media, including an emphasis on the letter as the vehicle for an idealized conversation and on its function as a means for the dissemination of knowledge, "di continuo aggiornamento, di informazione, di elaborazioni di opinioni" [of consistent updating, information, and elaboration of opinions] (Mori 20). Conversational salons constituted, according to Maria Teresa Mori, the most typical expression of nineteenth-century salon life in Italy, lasting for less than a century, from "la Restaurazione agli anni Settanta, con una particolare concentrazione nel periodo centrale dell'Ottocento prima e subito dopo l'Unità" [the Restoration to the 1870s, with a particular concentration in the central period of the nineteenth century and immediately after the unification] (15). Simonetta Soldani identified World War I as the official end of the "civiltà del salotto" ("Salotti dell'Ottocento" 557); though already in the 1880s, with the death of Claretta Maffei (and the closing of her famous salon in Milan), we can find the beginning of the decline of the "salotto dell'Ottocento," which was being superseded and amply replaced by newspapers as the new centres for the promotion and circulation of ideas and the production of public opinion.[38] An essential *trait d'union* between these two cultural traditions of sociability was letter writing. To write a letter in order to continue, elaborate on, or conclude a conversation conducted at a soirée of a prestigious salon was a common practice in the life of those who attended literary salons, and in newspapers the same cultural practice allowed journalists to reinforce the idea of the newspaper article as a cultural event (with its specific attributes in terms of time, space, and attendees), to foster a sense of community, and to communicate with a circle of cultural affiliates – the loyal readers and subscribers of the periodical – larger than any salon could reach. The growth of letters appearing in newspapers in the post-unification period constitutes an element of innovation but also continuity between past and present forms of cultural sociability, and is an expression of the mediating role played by nineteenth-century culture. As Janet Altman illustrates, "because of its 'both-and,' 'either-or' nature, the letter is an extremely flexible tool in the hands of the epistolary author," and one that facilitated the production of a cultural discourse which straddled the gulf between past and present, satisfying both the desire to remain rooted in the literary past and the urge to embrace modern forms of cultural expression (*Epistolarity* 43).

Letters began to appear in Italian newspapers towards the end of the eighteenth century – following the lead mainly of France and England, where this tradition was already consolidated (Cook 16–17). Not only periodicals devoted to a specialized female readership made the letter a central feature; mainstream newspapers also adopted the letter when attempting to target a larger public of readers.[39] The focus of this section, however, is on periodicals addressed to women or featuring a column (of letter-articles) directed mainly to the female members of the family, because it is here that periodical readership overlaps with the audience for popular fiction, including the novels analysed in the next two chapters. As I will illustrate, Verga and Serao became known to their readers first in the pages of these periodicals, and their readers approached the authors' epistolary fiction with an already established understanding of and familiarity with letter writing, including its literary properties and conventions. Both Verga's *Storia di una capinera* and Serao's sentimental fiction became particularly popular among female readers.

One of the first examples of letter use in Italian newspapers may be found in 1770, with *La Toelette*, a Florentine monthly considered to be the first periodical devoted to a female readership, which opened its twelve issues with letters, two of which were addressed to the Venetian writer Elisabetta Caminer.[40] Written probably by Giuseppe Pelli Bencivenni, or possibly his secretary, Giuliano Merlini, these letters were all thematically focused on questions related to women's education and role in society – topics that were to become central in many letter-articles published later in the nineteenth century. Issue number seven opened, for instance, with a letter to Elisabetta Caminer – prominent female journalist and editor of the prestigious *Europa Letteraria* and *Giornale Enciclopedico* – in which the author advocated a more conspicuous presence of books in women's life as a source of companionship: "Nelle attuali circostanze altri maestri non possono avere le femmine che i libri, compagni fedeli, amici senza riguardi, precettori illuminati e sinceri servono nella felicità, nella gioventù e nella vecchiaia" [In the current circumstances, women cannot have teachers other than books, loyal companions in youth and old age].[41] Inspired by the ideals of the Enlightenment,[42] the debate on female education assumed here the tone of a moral mission seeking to transform society by exposing women to inspirational readings that were serious but not intimidating – "massime solide giocondamente rivestite di una abito dilettevole" [serious matter playfully dressed in a delightful habit].[43] By promoting educational and entertaining content, these articles aimed to emancipate women without, however, adopting radical positions in terms of women's legal rights or social roles between the sexes. These articles presented a modality of female behaviour that was indissolubly connected to women's maternal role – a role rooted in the notion that motherhood was the first and primary social connection to which humans were exposed after birth. "La Repubblica," one of the two letters addressed to Caminer said, "dovrebbe interessarsi ad allevarle [le donne] meglio di noi, perché sono le prime le quali c'insinuano col latte quelle idee che di tutte le secondarie idee sono il primo anello" [The Republic should be interested in raising them better than we do, because they are the first to feed us with milk those ideas which are the main link to any secondary ideas].[44] Such a notion of motherhood anticipated the principle of women's educational mission both within the family and in society in general, and became an axiomatic truth in the nineteenth-century bourgeois ideological and cultural affirmation of the family.

La Toelette lasted only a year; soon after its twelfth issue was published in June 1771, the periodical was forced to close under pressure from competing publications, as the editor bitterly admitted in his farewell letter to Elisabetta Caminer, complaining about the publication of hundreds of novels "i quali sono per lo più una scuola del vizio nascosto sotto belle apparenze" [which are most of all a school of vice hidden under beautiful semblance].[45] Despite its commercial failure, *La Toelette* represented a remarkable journalistic experiment, anticipating some of the key elements, including the use of epistolary writing, that shaped the format and content of later periodicals. The letter proved to be an effective strategy of communication, and not surprisingly many nineteenth-century intellectuals adopted it when writing articles that were supposed to be simultaneously educational and entertaining, for these two elements made the text more easily comprehensible and approachable for the emerging public of female readers, newly literate but still lacking the classical training that would allow them to engage with more complex texts.

The letter featured prominently in one of the major and long-lasting fashion periodicals of the nineteenth century: *Il Corriere delle Dame.* As Silvia Franchini noted, fashion in the context of this publication constituted a main indicator not only of fashion trends but, perhaps more importantly, of changes in social behaviour. "La moda divenne un'arte che," wrote Franchini, "doveva suggerire il legame con la sfera dei sentimenti, stabilendo una connessione tra l'essere e l'apparire; al tempo stesso funzionava da indicatore raffinato della collocazione dell'individuo all'interno delle nuove gerarchie sociali" [Fashion became an art which suggested a link with the sphere of sentiments, making a connection between being and appearing; at the same time it functioned as a sophisticated indicator of where the individual stood in terms of social hierarchies] (39). Cause and effect of this educational mandate, the letter became a main instrument in *Il Corriere delle Dame* for the diffusion of its ideas and cultural projects.

Founded in Milan in 1804 by Carolina Arienti and her husband, Giuseppe Lattanzi, by 1811 it had a considerable number of readers and subscribers, and for almost a century it enjoyed a previously unprecedented commercial success.[46] Under the later, successful ownership of Giuditta Lampugnani and her son Alessandro, in 1848 the periodical began publishing a bi-weekly supplement, *La Ricamatrice* (renamed *Giornale delle Famiglie. La Ricamatrice* in 1860), which besides offering fashion patterns and embroidery designs provided readers with a literary section containing short stories, book reviews, and articles on various topics of social and cultural interest. Initially distributed only in northern Italy (in 1851, for instance, Giuditta Lampugnani regretted in a response to a reader's query being unable to send her periodical to Greece and Tuscany), from the middle of the 1860s it expanded its circulation to the southern provinces as well – an expansion evidenced by the appearance of the names of southern cities in the readers' letters included in *Corrispondenze.*

One of the sources of Lampugnani's periodicals' enduring success has been identified by critics as the original and skilful inclusion of letters, perceived at first as a novelty and later regarded as an essential feature (Franchini 186–7). The letter rubric had already been a feature of *Il Corriere delle Dame* under the ownership/editorship of Carolina Arienti, although until 1849 it mainly consisted of brief communications with readers on practical issues dealing with mishandled deliveries, or decisions on how to decorate one's house or on what outfit to wear for an event. Once Lampugnani took over, however, the rubric began to include articles, letters from readers, as well as short stories and poems written by prominent intellectual figures and journalists such as Francesco Dall'Ongaro, Caterina Percoto, Carlo Tenca, Pacifico Valussi, Arnaldo Fusinato, and Erminia Fuà Fusinato. Not all the letters sent by readers were published, of course; only a few of them made it into print – but the editor paid attention to them. As Giuditta Lampugnani admitted in 1855 in *Corrispondenze,* "Noi amiamo anzi codesto scambio di corrispondenze che può sempre guidarci al meglio, e procuriamo, per quanto ci è possible, di mantenere quella vita di azione e reazione col pubblico a cui dobbiamo in gran parte la buona accoglienza del giornale. Corrispondere ai desideri delle associate è certo il miglior mezzo di assicurarsene la simpatia" [We love this exchange of correspondence, which can guide us towards improvement, and try, as much as possible, to keep that relation of action and reaction with the public to whom we owe the good reception of the newspaper. To correspond to the subscribers' desire is the best way to ingratiate ourselves with them].[47]

Readers valued this space of communication, where they could keep up with the latest fashions and ideas and at the same time express their concerns and feel engaged in cultural activities. Given that venues for intellectual exchange such as academies or cafés were normally not available to Italian women, the space for correspondence provided by the newspaper represented a rare opportunity for female readers to be engaged in a conversation that took them virtually outside their domestic realm but physically allowed them to remain safely inside it. Not surprisingly, women letter writers preferred to remain anonymous, and editors complained about this, deploring the fact that anonymity fostered feeble argumentation and lack of accountability.[48] At a time when the national rhetoric of the family emphasized the domestic role of women, whose freedom of movement within the public realm even in urban settings was quite limited, the newspaper with its correspondence feature offered a space in which to convene regularly and feel part of a debate that involved a community of female readers, albeit a small one.[49] Far from being a mere outpouring of personal concerns, the topics of these letter exchanges ranged from embroidery to music and literature. In the *Corrispondenze* of 16 January 1855, for instance, Giuditta Lampugnani reassured those readers who, though praising in their letters the importance of music ("ché l'Italia è la patria di questa bell'arte, e non deve umiliarsi a copiar dai Francesi" [since Italy is the land of origin of this beautiful art and should not be humiliated by copying the French]), feared that too many articles on this topic would overshadow other important fields of discussion. "Le otto pagine di walzer dell'ultimo numero," wrote Lampugnani, "devono bastare pel momento a soddisfare quel vostro amore di patria onde vorreste approntato da capo a fondo il femmineo giornale" [The eight pages of waltz in the last issue should be sufficient for the moment to satisfy your love for our country with which you would like our female magazine to be filled].[50]

When Alessandro Lampugnani was solicited by readers to increase the space devoted to the publication of letters, he responded on 1 August 1862 by complaining that it was a recent decline in subscriptions that had resulted in a decrease of published letters. "Ancor oggi la Ricamatrice ha la metà delle associate che aveva prima del '59; e sta bene che queste superstiti non abbiano molta lena di far chiacchiere, non foss'altro per dolore d'aver perduto le compagne, fondamento di ricchezza pel giornale e fonte di belle tavole e di piacevoli articoli" [Today La Ricamatrice has half the number of members it had in 1859; it's all well and good that these survivors don't feel like chatting if nothing else for the pain of having lost their friends, the foundation of the periodical's richness and the source of beautiful tables and entertaining articles].[51] Regardless of the commercial vulnerability of the periodical – dependent for its survival on the number of subscriptions – *Corrispondenze* had by then become a familiar and expected feature, and readers considered it an essential component of the periodical.

Around the middle of the 1850s instructions on how to write letters also began to appear. In 1854 and in 1855, Lampugnani devoted four long articles in *La Ricamatrice* to the topic of letter writing. The first two, titled "Dello stile epistolare I" [On epistolary style I] and "Dello stile epistolare II" [On epistolary style II], provided general rules of letter writing, elaborating on the importance of finding one's own writing style "come ognuno ha un carattere calligrafico diverso, così ognuno ha un carattere morale diverso" [as we all have a different writing style, so each of us has a different moral character], and emphasizing the need to modernize one's writing skills in light of the current changing times, "senza perderci in tutte le minuzie di quel codice epistolare tenuto in tanta reverenza da' nostri padri, ma che i tempi mutati ci comandano di considerare oggimai senza rimorso come lettera morta" [without getting lost in the minutiae of the old epistolary code so revered by our fathers, but made obsolete by our changed times, which force us to consider it, without remorse, a relic of the distant past].[52] As with the guidelines that Marchesa Colombi and Emilia Nevers would later provide in their conduct books, the anonymous author of these articles underscored the need to simplify and minimize epistolary formalities, assuming that women would mostly write familiar letters for which they would use natural and simple forms of expression derived from everyday conversation. However, these epistolary instructions were not just about stylistic and linguistic propriety; they also warned about the risk of exposing oneself to social scrutiny, and possibly scorn, through a poor choice of language. *Verba volant, scripta manent*. The Latin saying resonated deeply in this article, which ended with a reminder that moderation should be the guiding principle of letter writing – a principle that conflicted with the spontaneity correspondents were also encouraged to capture. "Quella moderazione e quella dignità e quella prudenza che devono informare le parole che volano, sono tanto più da inculcarsi negli scritti che restano. Quante volte si pagherebbe di proprio sangue l'imprudenza d'una parola, e quante volte una parola scritta costò appunto e sangue e vita" [That moderation, dignity and prudence which must inform words that fly away must be even more instilled in writings that have permanence. How many times have we paid in blood for one imprudent word, and how many times has a written word cost us blood and life?]. The following year, in the October and November issues, two long and detailed review articles were published on Giuseppe Picci's *Guida allo studio delle belle lettere ed al comporre con un manuale dello stile epistolare*. These lengthy descriptions of epistolary rules represented neither an advertisement for Picci's book nor a critical evaluation of the manual; rather, they offered a useful synopsis of the book's content for a readership that, having more leisure time and therefore more time for correspondence, needed, according to Lampugnani, guidance in mastering this skill ("argomento, siccome di spettanza altrettanto femminile che maschile" [a topic that is useful to both men and women]).[53] A savvy publisher and editor, Lampugnani strove, through an alchemic combination of instructive material and entertaining readings, to produce a publication that appealed to the different needs and expectations of readers, so that, as in the case of Picci's manual, a text originally intended for a relatively small group of readers – mainly male high-school students – was brought to the attention of a much wider audience – both women and men – of different ages and social classes.

During the 1850s and 1860s, letter-articles published in *La Ricamatrice* (renamed in 1860 *Giornale delle Famiglie. La Ricamatrice*), considered by critics a "veicolo linguistico d'italianità" (linguistic vehicle of Italianness), intensified the debate over national reforms (Franchini 191). Intended to promote social and moral progress, these articles often focused on what in the nineteenth century was referred to as *la questione della donna*, the question of woman's social, legal, and cultural role in society. Lampugnani's periodical, though moderate in tone, offered articles that were meant to influence public opinion and raise awareness among the reading public of the importance of female education for the moral development and social advancement of Italy. On 16 April 1852, for instance, a letter-article with the title "Francese o Italiana?" [French or Italian?] lamented the lack of knowledge of the Italian language among young Italian women, who mainly spoke either French or a dialect: "Voi trovate in Italia fanciulle le quali, a quindici anni sapranno scrivere un bigliettino in lingua francese profumatissimo, e non saprebbero raccontarvi in buon italiano la favola del Corvo e della

Volpe" [You will find in Italy girls who at fifteen can write a perfect note in French but would not be able to tell you in good Italian the tale of the Crow and the Fox].[54] Similar concerns were expressed in *Corrispondenze* in 1859, in another letter-article titled "Sulla convenienza per le fanciulle di adoperare anche nell'uso domestico la lingua comune italiana" [On the convenience of young women using the Italian language in the domestic realm], where Ippolito Nievo, under the pseudonym "N.N.,"[55] provocatively asked the recipient of the letter, *signora Teresa B.* from Mantua, whether she would be willing to persuade "le nostre giovani compaesane della necessità di abbandonare anche nei colloqui famigliari l'usanza del rozzo e vario dialetto per quella più nobile e sicura della lingua comune" [the young ladies of our country of the need to abandon even in familiar conversation the custom of using the crude and varied dialect in favour of the more noble and secure common Italian language] (see the full text in the Appendix).[56] For Nievo, women of the upper classes had to speak Italian at home, not only for their own edification, but also for that of the lower classes. If servants, he thought, were addressed more frequently in Italian they would begin not only to understand it more, but to use it. From February to March of 1857, the newspaper published a series of five letters titled "Donna italiana," addressed to Caterina Percoto and written by Pacifico Valussi, renowned journalist and director of *Perseveranza*. In these letters, Valussi used the rhetorical tool of the *captatio benevolentiae*, a common practice inherited from the medieval tradition of epistolary conventions and aimed at reaching a wider audience through the intercession of a public figure, Percoto, who represented an intellectual authority among readers. Valussi admitted in his first letter that he needed "che voi donna m'introduciate nella società del vostro sesso, cattivandomi presso le altre un po' di quella benevolenza che mi accordate. Ed eccovi tutto il segreto della lettera, che per venire a voi, a poca distanza da Udine, prende la via di Milano e si mette nelle colonne della *Ricamatrice*" [you as a woman to introduce me into the society of your sex, instilling in them some of the benevolence you feel for me. This is the full secret of this letter, which to reach you outside Udine has to travel by way of Milan and appear in the columns of *Ricamatrice*] (see the full text in the Appendix).[57] Dwelling at length in the five letters on the importance of female education in equipping women for their social role as mothers – "La donna adunque è essenzialmente madre; è educatrice prima di tutto […] di qui la necessità per lei di rafforzare il suo affettuoso istinto con appropriata educazione, di rendersi insomma colta a fungere l'ufficio di madre" [Woman is therefore essentially a mother, she is first of all an educator ... from this follows the need for her to strengthen her affectionate instinct with an appropriate education, to become in other words knowledgeable about how to play her role of mother] – he concludes with praise of motherhood and proposes the creation of an *enciclopedia delle madri* [encyclopedia of mothers] – an idealistic compendium of works written by women for the instruction of other women (Letter V; see the Appendix).[58] With this idea in mind, Valussi advocated the production of a *letteratura femminile* – "studi che alle donne medesime competono" [studies which are the competence of women] – which, with its educational content, would give women "tanto della classe colta, come della media e povera, i mezzi per essere, oltreché madri, educatrici" [of the educated as well as middle and lower classes the means to be educators in addition to being mothers].[59] The topic of education continued to occupy a central space in the periodical, as articles explored the different implications of school reforms and social and gender identities. In 1861, for instance, it published two articles titled "Pensieri sull'educazione: Corrispondenza di una signora," comparing the educational and social reforms of Italy and England; while in 1864, "Lettere sull'istruzione," written by an unidentified "family's father" discussed the need to implement reforms not only at the primary and secondary but also at the university level. During the same year, several letter-articles were published on the topic of forcing young women to enter a convent, exposing the shortcomings of women's education in religious institutions (Caterina Percoto wrote four of these articles, which will be discussed in the following chapter in the context of Verga's novel). The educational debate continued to occupy a prominent space in the periodical, involving a range of voices, some male some female, some well known (Nievo,Valussi, and Percoto), others less so, who contributed to producing public opinion among the still numerically limited but growing public of female readers.

Another periodical to use the letter regularly in the second half of the nineteenth century was *Passatempo,* founded in Turin in 1869 by Amerigo Vespucci (sic). A monthly until 1872, it was then renamed *Giornale delle Donne* and published biweekly until 1940. Along with *La Donna* publishedby Gualberta Alaide Beccari, *Giornale delle Donne* was more outspoken than most newspapers in advocating for the social and cultural advancement of Italian women. Among its objectives, as stated in the paper's first issue, was that of "opporsi alle mollezze d'oltr'Alpi con morali, dilettevoli ed istruttive pubblicazioni" [opposing the feebleness coming from beyond the Alps [France] with moral, entertaining and instructive publications]. The founding idea was, thus, to create a paper with a modern approach to journalism and a specific ideological agenda, in which didactic purposes complemented commercial objectives of entertainment. It was a matter of responding to what was perceived as a national call for the cultural and spiritual awakening of Italian women, but also of competing against other periodicals that vied for readers' attention and subscriptions.[60]

Readers reacted immediately to Vespucci's opening mission statement. In the second issue, the editor published several letters he had received from his readers. One letter, from a woman who lived in Veneto, praised the format and content of the paper because "se v'ha un bisogno urgente ed oltremodo urgente per questa nostra patria è quello di dare un indirizzo utile e nuovo affatto all'educazione della donna" [if there is a single and most urgent need for our country it is that of giving a useful and altogether new direction to women's education].[61] The letter appeared in a column titled *Conversazioni in famiglia,* and later renamed *Divagazioni* and *Rose e Spine.*

The letters included in these columns were written by ordinary citizens, anonymous correspondents, and aspiring authors – some of whom later became well-known literary figures – as well as already established writers. In the 1873 August issue, for example, *Conversazioni in famiglia* contained letters from two readers, Lina Baroni and Baronessa Irene De-Morand, who were actively encouraging their friends to subscribe to the paper so that "in tutte le famiglie della nostra cara patria si seguano le massime che il suo giornale impartisce alle donne italiane" [in all the families of our dear country, the ideas about Italian women that your newspaper teaches may be shared]. *Conversazioni in famiglia* was featured at the end of each issue and provided a space devoted to correspondence with readers in which the director, Vespucci, quoted, replied to, but also debated the recommendations and criticism he received from subscribers. Most of these correspondents were cited by name, but a few still opted to remain anonymous, as in the case of a reader from Palermo who, presented as "Signora C.," chastised Vespucci for not reviewing Serao's *Romanzo della fanciulla,* an accusation from which he defended himself by lamenting the publisher's lack of initiative in sending him a copy to review. But it was not only readers who wrote letters; writers did, too. Neera, Marchesa Colombi, Matilde Serao, Tommasina Guidi, Erminia Fuà Fusinato, and Emilia Nevers, to cite some of the most famous female authors of the time, submitted essays and articles in letter form, often addressed to each other. In 1870, for example, Emilia Fuà Fusinato, educator and poet, wrote a series of seven letter-articles for *Passatempo* on the topic of female education.[62] While formally addressed to the director of the paper, Amerigo Vespucci, these articles were meant to engage the larger public of female readers in a conversation ("come se si conversasse tra noi") [as if we were conversing among ourselves] on the topic of women's emancipation, which for Fusinato could be attained only through education and a gradual liberation from obsolete customs and atavistic ignorance. In 1873, Caterina Pigorini Beri wrote three letter-articles similar in content and tone to Fusinato's, discussing the current state of the newly created *scuole normali femminili* [high schools for women]. These articles, describing the administrative and pedagogical disarray in which these schools were forced to operate, were also republished in other newspapers and provoked a lively debate among and beyond the readership of *Giornale delle Donne* (by then the new name for *Passatempo*).

Mainstream newspapers also made ample use of letter-articles, and invited women's writers to publish articles that would appeal to their female readers. In *L'Illustrazione Italiana*, for instance, on 16 April 1876, Marchesa Colombi's letter-article "La donna povera" [The poor woman] responded to the article "La donna libera" [The free woman] by Neera published in the same newspaper two weeks earlier, on 2 April. Condemning the notion of female employment, and following the principle "togliete la donna alla casa, e non avrete più né casa né donna" [take the woman out of the home and you will have no longer either a home or a woman], Neera defended her position by adducing a moral argument in favour of a traditional division of gender roles so as not to increase the already high level of men's unemployment.[63] Marchesa Colombi, ideologically more progressive than Neera, responded by focusing her argument on women's need to find a job, especially middle-class women, who, often unable to secure a good dowry and financial support from their families, saw employment as the only way to avoid poverty.[64] (The full text of both letters is provided in the Appendix.) A similar discussion was carried in *La Lega della Democrazia*, a progressive newspaper founded in Rome by Giuseppe Garibaldi, Alberto Mario, and Agostino Bertani, which on 15 May 1880 published the emancipationist Anna Maria Mozzoni's "Lettera aperta a Matilde Serao" [Open letter to Matilde Serao], in which she criticized the Neapolitan writer's view that women did not have political opinions. Two weeks earlier, Serao had written an article to Martino Cafiero, the director of *Il Corriere del Mattino,* in which she had explained what she meant by women's "politica del cuore," an expression she used to define what she considered women's natural indifference towards politics.[65] Mozzoni, referring to the political work of women around the world in fighting for "abolizione della schiavitù,' nella riforma penitenziaria, nel richiamo degli atti sulle malattie contagiose e nella riforma delle loro condizioni giuridiche e sociali" [the abolition of slavery, prison reform, [and] in their efforts to cure contagious diseases and reform their social conditions and legal status], exhorted Serao to expand her horizon of knowledge about what was happening outside Italy and to correct her view of women's natural attitudes.[66] Serao replied a week later, on 21 May, with an uncompromising argument in which she emphasized that her article was based on an artistic rather than a political point of view. Serao's rejection of the figure of the emancipated woman was presented as part of an aesthetic judgment that perceived the modern, emancipated woman as fundamentally stripped of beauty and poetry. (The full text of the letters may be found in the Appendix.) Finally, mention should be made of another correspondence, published in *La Stampa,* in which Matilde Serao debated with Marchesa Colombi about housekeeping and servitude. In two articles, published on 25 February and 1 March 1905, respectively, Serao disparaged and Marchesa Colombi defended the figure of the female domestic worker, presented by the former as being inherently servile and untrustworthy and by the latter as a victim of employers' insensitivity if not outright persecution.[67] Marchesa Colombi's article "La padrona" was written in the epistolary format, not only to respond personally to Serao's provocative criticism of domestic workers but to encourage readers themselves to join the discussion and ponder an issue – class mobility and social protest – that was filling the pages of all newspapers across the spectrum of conservative and radical journalism. In a similar vein, Marchesa Colombi's and Neera's letter exchange implicitly invited readers to form an opinion about the then much-discussed topic of women's employment and entry into the national labour force – whether as teachers or as labourers. Written in the intimate tone of a conversation exchanged between two interlocutors of opposite ideological views, these letters assumed the status of public speech designed to influence readers. The letter form provided these female intellectuals with a formidable tool for communicating their arguments and effectively challenged their readers to join the public debate.

In conclusion, the emergence of letter writing in post-unification Italy reflected a thriving cultural and social practice, comprising private as well as public correspondence, educational projects, reformist programs, and cultural and social pamphleteering, all framed within national debates regarding the definition of Italy's new cultural, social, and gender identities. The letter's Janus-like character (nostalgically glancing backward towards tradition and yet eagerly looking forward towards modernity by way of technological innovations) became instrumental in the expression and circulation of ideas, but was also emblematic of Italy's ambiguous relationship with cultural modernization. Reading and writing letters was an accepted and even encouraged practice for women as long as it was kept within the confines of a moderate public debate (often controlled by a male figure). Ambiguities may also be found in the epistolary fiction of the post-unification period. Verga and Serao, whose work is analysed in the next two chapters, adopted for their fiction a genre that was becoming outmoded in the rest of Europe but still afforded them with an effective literary strategy for engaging the growing public of readers and experimenting with a "modern" approach to literary production – an experiment that was necessary for any writer who aspired to participate in the development of a modern Italian literary tradition.

PART II

3 Fictionalizing the Letter: Giovanni Verga's *Storia di una capinera*

... un giorno viene da me una persona che conosco e che mi fa leggere una lettera ... una lettera molto strana ... Che magnifico punto di partenza per un romanzo!

[... one day a person I don't know comes to me and makes me read a letter ... a very strange letter ... What a magnificent point of departure for a novel!]

Neera[1]

Written in a moment of intense nostalgia for his family while he lived in Florence in 1869, "una notte che passeggiava, solo, lung'Arno, evocando i ricordi del tetto paterno" [during a night walk, by himself, along the Arno, while he recalled the memories of his father's house], as De Roberto famously recalled, *Storia di una capinera* is the novel that inaugurated Giovanni Verga's debut on the national literary scene and brought him to national fame (*Casa Verga* 142). Its publication was greeted with general enthusiasm: readers viewed it as a true novelty of the burgeoning literary market, and critics hailed it as a work of fiction full of promise for the renewal of the national literary tradition.[2] In 1873, when the novel's second edition came out in Milan with the publisher Treves, the many positive reviews by some of the most renowned intellectual figures of the time – Angelo De Gubernatis, Francesco Dall'Ongaro, Ferdinando Martini, Salvatore Farina – testified to the critical success of a novel relegated today to a seemingly opaque phase in Verga's career and generally disparaged for its lack of real literary merit.[3] Although some critics, like Musumarra, recognized that Verga's epistolary novel constituted an initial narrative search for authorial objectivity, later fully expressed through his *verismo* (68), and others pointed to the importance of this work from a linguistic and stylistic point of view (Brusadin, introduction to Verga, *Storia* xv), *Storia di una capinera* is usually neglected by critics and marginalized within the literary canon.[4]

In light of this negative critical perception, my reading of Verga's *Storia di una capinera* aims at returning some of the critical attention to the historical and cultural circumstances that determined the original success of this work, a novel that, however imperfect it may have been, Verga never repudiated and that marked a crucial point in the author's career. We may also consider the role this novel played in the formation of Italy's post-unification literary identity – what Giovanni Ragone defined as the "passaggio ad una cultura moderna," that is, the post-unification development of a modern cultural identity and literary sensibility (*Un secolo di libri* 3). Strongly interconnected to contemporary literary debates and cultural practices, *Storia di una capinera* provides useful insights into our understanding not only of the author's early literary career but also of a historical period in which cultural and institutional interventions began to produce a national consciousness and a new modern literary identity. Such a synchronic approach is also useful in illustrating the interplay among the different agents involved in the process of production and consumption of this novel – Darnton's communication circuit, mentioned in the previous chapter – thus offering new interpretive opportunities for *Storia di una capinera* that enable us to see it as the result of a complex and layered nexus of cultural interactions rather than merely a failed attempt to fulfil authorial artistic aspirations. As Darnton explains, a text – whether a Shakespearian sonnet or a bicycle manual – is necessarily informed by a communication circuit, which runs from the author to the publisher and the reader. This is not to say, of course, that a sonnet and a manual are textually one and the same, but rather that the literary text, too, is bound to follow a communicative mechanism that implies an interaction between a sender, a transmitter, and a receiver of the communication – a multiple agency circularly interconnected in the process of information exchange. "The reader completes the circuit because he influences the author both before and after the act of composition. Authors are readers themselves. By reading and associating with other readers and writers, they form notions of genre and style and a general sense of the literary enterprise" (Darnton 11). The interaction among the different agents is what interests me most in my analysis of Verga's epistolary novel, and I view the letter as a crucial instrument in fostering this exchange of information through which a sense of belonging (to a community of writers and readers) is created by way of engaging both reader and writer in a shared interpretive and communicative experience. Critics have shown that the epistolary form facilitates the production of associations between reader and writer because it is "unique among first-person forms in its aptitude for portraying the experience of reading [...] making the reader (narratee) almost as important an agent in the narrative as the writer (narrator)" (Altman, *Epistolarity* 88). My analysis of *Storia di una capinera* will therefore look both at the formal elements of the novel and at various aspects of the interrelation between text, author, and reader, thus focusing less on posthumous criticism on Verga and more on contemporary sources, such as the cultural debates, reviews of the novel, and established and emerging literary discourses, as well as then current practices of epistolary writing, in an attempt to investigate the cultural and social environment in which Verga operated as a main interlocutor in the national literary discourse and cultural interplay of post-unification Italy.

The transformation of the print media in nineteenth-century Italy brought to the fore a crucial question regarding the process of modernization of the cultural industry, that is, the relationship between production and consumption of the literary product, between author and reader. In the age of mass readership, beginning with the post-unification period, the reading public began to be perceived as "unfathomable and intimidating by virtue of its vertiginous expansion," and writers became increasingly and eerily aware of the demands that both readers and the publishing establishment were putting on them to produce works that, in addition to artistic merit, would appeal to a large audience and produce a substantial profit (Lolla 30). It was not just a matter of selling books, however, but of proving that, by having a sizeable national audience, both authors and publishers had a national appeal and contributed to the formation of modern literary and cultural identities. Now that political unification was achieved, the new nation of Italy was eager to prove itself equal to its European counterparts, at least culturally if not economically, and a strong emphasis was put on efforts aimed at developing a modern cultural profile in Italy. Verga's literary career exemplifies those efforts as well as the anxiety felt by many authors of the post-unification period as they sought to address the needs and expectations of the evolving publishing world. Unlike Carlo Tenca, who, in an article published in the *Rivista Europea* in 1846, complained about the effects produced by the process of cultural democratization, by which the writer, "emancipandosi dalla protezione dei grandi, e dipendente invece da un vasto popolo di lettori" [freed from patrons' protection, and dependent instead on that of the vast public of readers], inevitably becomes "servo dei capricci e della versatilità del pubblico" [slave to the caprices and versatility of the public] (92), Verga seemed to accept the role that the reading public played in the development of a writer's career. "Ciò che mi conforta a credere, meno qualche eccezione, che il primo critico dei suoi scritti si è l'autore stesso, e poi il gran pubblico" [This brings me to believe that, with a few exceptions, the first critic of someone's work is the author himself, and then the larger public], Verga wrote to Treves in 1873, eager to emphasize the positive reception by the public of his two novels, *Storia di una capinera* and *Eva*, and, thus, to affirm his rising status as a national literary figure (Raya, *Verga e i Treves*, 30). Treves himself had solicited critics to review Verga's novel. "Se tutti scrivono sarà un bel coro" [if they all write about it, it will be a nice choir], the publisher commented in a letter, hoping to spark interest, first of all among the professional readers who would be in the position of promoting the publication among the wider public of readers; "manderò io le copie ai giornali non solo, ma scrivo e riscrivo ai critici perché parlino, al caso male, secondo coscienza – ma parlino" [I will not only send the copies to the newspapers, I will also write and write again to critics so that they write about it, even negatively, according to their conscience, as long as they will write about it], insisted Treves (Raya, *Verga e i Treves*, 28). Both publisher and author grasped the relevance of what was then called *réclame*, the space in newspapers in which the publication of books was publicized, and of what critics could do for a young author eager to climb the ladder of literary fame. As one critic commented in a review in *Perseveranza*, referring to the relationship between the "seasoned" Treves and the still "young" Verga,

> Ci voleva un editore intraprendente e sagace, che scoprisse nel giovan e ignorato scrittore la stoffa del romanziere e gli desse l'aìre; ci volevano i mezzi di cui solo una grande Casa editrice può disporre, perché la gran massa del pubblico imparasse a conoscere le opere di Verga, le discutesse e le giudicasse. Forse i giudizi non sono concordi [...] ma che importa? La discussione è la vita, è la luce, è – per uno scrittore d'immaginazione specialmente – l'avvenire. La lode dei numerati amici non basta a compensare il silenzio del pubblico, il quale ha bisogno di essere stimolato, condotto quasi per forza a scegliere tra la moltitudine delle produzioni, che tuttodì escon dai

torchi, leggere e giudicare. Torni Dante, e se nessuno gli farà, come si dice, la réclame, passerà inosservato anche lui. (14 September 1873)

[There needed to be a shrewd and enterprising publisher, who would discover in the young and ignored writer the makings of a novelist and give him a boost; there needed to be the means that only a large publishing house could provide, so that the masses could learn to know Verga's works and to discuss and judge them. Perhaps the judgments would not be in agreement with one another [...] but what does that matter? The discussion is the life, the light, it is – especially for an imaginative writer – the future.

Praise from numbered friends cannot compensate for the silence of the public, which needs to be stimulated, and is almost forced to choose among the many productions in print today, to read and judge. Dante could be here today, but if no one gave him, how do you say, any publicity, he too would pass unnoticed.]

Verga, like Treves, understood the complexity of the dynamics regulating the modern world of literary production and consumption, a world that in post-unification Italy, too, regardless or perhaps in light of its industrial backwardness, could not afford to dismiss the readership as a marginal aspect of the cultural landscape. Torn, however, between the lure of commercial success and the ideal of creative independence, Giovanni Verga, like many other writers, demonstrated both an eagerness to reach out to readers but also an ambiguity about what was viewed then as the "democratizzazione dell'arte" [democratization of art],[5] an expression used with disdain by some intellectuals, who, as in the case of Edoardo Scarfoglio, comparing "gli operai della letteratura" [literary workers] to "le meretrici vagabonde" [wandering prostitutes], protested vehemently against the "imperio della maggioranza" [the rule of the majority] (207).[6] But if writers claimed artistic independence from the reading public, professionally they necessarily depended on it; modernity had brought opportunities to those who wanted to become professional writers, but also constraints, and with them the frustration at times of being part of a system that, moving from an artisanal mode to a more industrial one, complicated the relations between the parties involved in the production and consumption of the literary product.

The relationship between author and reader dominated the cultural and literary debates of the second half of the nineteenth century (Cadioli 121). Ruggero Bonghi, the founder of *La Stampa*, director of *Perseveranza* from 1866 to 1874, author, and member of the Italian parliament, was one of the first writers to raise publicly the question of how to make Italian literature (i.e., Italian novels) more popular among Italian readers.[7] From 18 March to 7 October 1855, Bonghi published in the Florentine *Spettatore* sixteen letters addressed to Celestino Bianchi, the newspaper's director, in which he lamented the lack of interest showed towards Italian authors:

> I libri italiani hanno in Italia un molto minor numero di lettori che i francesi in Francia, i tedeschi in Germania e gl'inglesi in Inghilterra. [...] E non è già che in Italia si legga assolutamente meno che altrove; si leggono meno i libri nostri, e a quel sovrappiù di lettori a cui i libri nostri o non bastano o non piacciono, suppliscono quelle altre tre letterature. (66)
>
> [Italian books have in Italy a smaller number of readers than the French ones have in France, the German ones have in Germany and the English ones in England. And not because in Italy people generally read less than in other places; our books are read less, and those additional readers who don't like or don't find our books sufficient are compensated by the other three literatures.]

Bonghi's letters were then compiled in a volume, published the following year and prefaced with a letter to Giulio Carcano, in which he argued in favour of modernizing the figure of the writer by renewing his relationship with the reading public – "un pubblico per un autore è un'ipotesi necessaria" [the public is a necessity for the author] (35) – and adopting a literary language closer to the one spoken by the readers:

> Sono poi convinto, che questa pompa e quest'artificio non potrebbero effettivamente diminuire, se alla lingua cortigiana, signorile, affettata, monca, sforzata, ipotetica che ora è adoperata dagli scrittori, non si surroghi quella viva, compita, spontanea, reale che è parlata davvero nella più gentile delle province italiane. (38–9)
>
> [I am also convinced that this pomp and artifice could be effectively diminished – this language now used by writers, formal, aristocratic, affected, truncated, forced, and hypothetical – if replaced by a new one which is alive, complete, spontaneous, real and which is really spoken in the most refined of the Italian provinces.]

La questione della lingua had been a major hurdle to the creation of a modern national literary tradition, as Manzoni had amply illustrated in both his theoretical and narrative writings. Bonghi's concerns were certainly not new, but his letters had the merit of bringing the issue to the attention of non-specialists. His book was republished twice, in 1873 and 1884 – the last edition being for schools – attesting to the relevance of a question still much debated and considered by some writers insoluble, at least at the time.[8]

While by 1895 several Italian authors had become popular among readers (De Amicis, Fogazzaro, and Serao were the most popular, according to a poll taken in 1906),[9] the discussion about how to balance the authors' need for national visibility in the publishing market with their need for artistic independence from consumer culture continued throughout and beyond the nineteenth century. An overview of this literary debate may be found in Ugo Ojetti's *Alla scoperta dei letterati* (1895), a collection of interviews with some of the main figures of the contemporary Italian literary world, "quei venti o trenta scrittori che o per verace valore d'opera o per sola fama o per questa e per quella hanno ormai messo il capestro al calcitrante asinello chiamato pubblico" [those twenty or thirty writers who either for true literary merit or for simple fame or for this or that have latched onto the wild donkey known as the public], to assess their opinion on the state of modern Italian literature (vii). A generational divide seemingly characterized the different positions expressed by the interviewed authors. If, on the one hand, Federico De Roberto, of the younger generation, suggested that writers should choose subjects of interest to readers, on the other hand, Cesare Cantù, of the older generation, complained that books were no longer created in accordance with the author's aesthetic taste but with that of the readers. And while Edmondo De Amicis advanced a functional and social mission for literature, under the aegis of a socialist ideology that assigned to the reader a militant role in the literary production of meaning ("credo che il romanzo sia la grande forza di battaglia e di propaganda [...] Ma non intendo che il romanzo abbia una tesi, il romanzo narrerà dei fatti coordinati a un'idea, ma la conclusione deve essere fatta dal lettore, non dall'autore" [I believe that the novel is a great force for battle and propaganda ... But I don't mean to say that the novel should have a thesis, the novel will tell facts coordinated to an idea, but the conclusions must be made by the reader, not the author]) (Ojetti 126–7), Gabriele D'Annunzio exalted the spiritual force of art, the absolute superiority of language above all other forms of human expression, because art, "che nelle sue forme supreme rimane godimento dei pochi, risponde in realtà a un bisogno diffuso" [which in its most supreme form remains the enjoyment of only a few, responds in reality to a diffuse need] (Ojetti 319). Finally, Giovanni Verga defended the novel as the best artistic vehicle to engage an ideal readership, since other artistic forms, such as the theatre, forced the author to write "per un pubblico radunato a folla così da dover pensare a una media di intelligenza e di gusto, a un *average reader*, come dicono gli inglesi. E questa media ha tutto fuori che gusto e intelligenza; e se un poco ne ha, è variabilissima col tempo e col luogo" [for a public gathered in a crowd so that one must think about an average intelligence and average taste, an average reader, as the English say. And this average has anything but taste and intelligence; and if it has a little of it, it varies greatly with time and place] (Ojetti 71). One might discern in Verga's response the sense of disappointment with the reading public he felt at that time; when Ojetti interviewed him in 1894, the Sicilian author had experienced the fickleness of a readership unwilling to follow him on his naturalistic literary venture. *I Malavoglia,* his *verista* masterpiece, had been a commercial flop, and Verga began to revise his view of the relationship between author and reader. But at the beginning of his career he viewed the reader, real rather than ideal, as a positive influence on his career as a writer.

When Verga moved to Florence in 1869, and later to Milan, he sought to be acknowledged as a writer for the artistic merit of his poetics. In a letter written on 24 July 1869, he told his mother that he longed to become a writer – an occupation, he remarked, that was highly regarded in Florence and potentially also profitable:

> Credimi, cara madre, che questo non è più il tempo che uno possa qui divertirsi, ma io m'impongo alacremente tutte le prove perché la speranza alfine di riuscire a qualche cosa di utile e di serio mi fa lavorare con accanimento e soffrire quasi con piacere tutte le privazioni (la maggiore delle quali quella della famiglia) pensando che in quest'inverno ne sarò compensato con bei guadagni, e con una posizione che qui è invidiabile perché assai rispettata. (*Lettere sparse* 39)
>
> [Believe me, dear mother, this is no longer the time for one to enjoy oneself here, but I tirelessly put myself through many tests because I hope to reach in the end something helpful and serious which makes me work with tenacity, and I suffer many deprivations (the main one being the family) almost with pleasure, thinking that next winter I will be compensated with good earnings and a position that here is envied because highly respected.]

But he also relied on the publishers' commercial initiatives to gain attention and approval from readers. The publication history of *Storia di una capinera* is, in this sense, revealing: it was published twice in instalments, first in Alessandro Lampugnani's *Il Corriere delle Dame*, between 16 May and 22 August of 1870, and then in Emilio Treves's *Illustrazione Popolare* between 9 March and 29 June 1873; and later printed in book form, with Lampugnani in 1871 and with Treves in 1873, 1883, and 1893.[10] Its publication, first in serial form and only later as a book, indicates a marketing strategy on the part of both author and publisher that aimed at maximizing public attention. The novel "piace anche sbonconcellata, e piacerà moltissimo unita" [is also liked in instalments, and will be very well received as a whole] announced Dall'Ongaro enthusiastically, after placing Verga's novel with Lampugnani (De Roberto, *Casa Verga*, 165). Verga was not ecstatic about the way in which his book was going to be published, as he mentioned in a letter to his mother, but viewed it as a necessary step towards literary recognition and economic advancement:

> Vi confesso che da un canto mi rincresce sciupare il mio nome in piccole pubblicazioni e di poco conto per le riviste e i giornali. E che vorrei riserbarmi intero, anche al prestigio di un nome acquistatomi coll'Eva [...] ma intanto quelle piccole pubblicazioni son quelle che fruttano dippiù. (*Lettere sparse* 52)
>
> [I confess that on one side I am sorry to waste my name in small publications of little merit, such as newspapers and periodicals. And I would like to keep myself whole, also in light of the prestige that Eva gave to my name ... but in the meantime those little publications are the most profitable ones.]

The book had initially been offered to Emilio Treves in Milan and to the Florentine periodical *Nuova Antologia*, but as Dall'Ongaro – who was interceding on Verga's behalf with the publishers – explained in a letter to the Sicilian writer, that initial plan did not go through; instead he had managed to make an agreement with Lampugnani to publish it in *Il Corriere delle Dame*, an established women's fashion magazine that featured a literary section and enjoyed a robust readership and a good reputation as an instrument of moral reform. Despite growing complaints about the negative effects of journalism on literature ("il giornalismo quotidiano che obbliga lo scrittore a scrivere presto e a scrivere in modo facilmente intelligibile a tutti" [the daily journalism which forces the writer to write quickly and in a way that is easily intelligible to all]), periodicals remained the main portal for writers who wanted to gain public exposure and reach a wider audience (Ojetti 224). Lampugnani, Treves, and Dall'Ongaro all understood that *Storia di una capinera*, given its fragmented epistolary format, was well suited to serialization and that its narrative energy would create a sense of eager expectation among readers that could increase the sales of the periodical.[11] Letters, with their enclosed and yet open-ended format, enticed the reader to continue reading the novel and therefore purchase the periodical. Both thematically and structurally, Verga's epistolary novel was well suited to create the sort of wait-and-see tension on which serialized publications relied for success. Furthermore, publishers, as discussed in the previous chapter, relied on readers' familiarity with and appreciation of letter writing, and assumed that a novel written in the letter form and thematically centred on the sorrows of a girl forced into a convent would grab readers' attention. Lampugnani's initial business intuition proved canny, although the novel's first edition of 1871 was, in terms of sales and circulation, slightly disappointing – "del libro si parlò poco" [they did not talk much about the book] lamented Verga (*Lettere sparse*, 102). By 1906 according to Treves – perceived by contemporaries as "uno dei migliori indici delle simpatie letterarie del pubblico colto" [one of the best predictors of the literary taste of educated readers] – *Storia di una capinera* had sold 20,000 copies, and it continued to sell very well (Tortorelli 159–60).

The commercial success of Verga's epistolary novel may be explained as the result of an alchemic convergence of several factors, from the author's artistic literary experimentation with established genres (by adopting specific linguistic and stylistic choices), to the publisher's cunning marketing efforts, and to the readers' reception of a novel interpreted in light of contemporary cultural debates and social practices (they themselves used letters, and what they read in letter form resonated with their own reality). Structurally presented in letter form and thematically based on the life of women in convents, *Storia di una capinera* captured readers' imaginations and marked a turning point in Verga's career.

Storia di una capinera tells the story of Maria, a young Sicilian woman forced into a convent by her family when her father, a widower, remarries and has to ensure that a dowry (mainly obtained through his second wife's estate) is made available for his two other daughters, Giuditta and Gigi, born of his second marriage. The novel comprises letters written by Maria to her friend Marianna, spanning a period of approximately two years, between 3 September 1854 and 24 September 1856, in which the protagonist confides to her friend the physical and psychological torments caused by her having to take the veil and renounce her love for a young man, Nino, who will eventually marry her stepsister, Giuditta. The tragic death of Maria, announced in a final letter written to Marianna by a sympathetic nun, Suor Filomena, concludes a narrative not uncommonly described by contemporary and later critics as belonging to the genre of the pathetic and sentimental ("monocorde cronaca patetica" [a monotonous and pathetic chronicle], according to Gino Tellini, and "tono da patetico in lezioso e sentimentalistico" [a tone that goes from pathetic to affected and sentimental] for Maura Brusadin) – with a pathos developed thematically through the figure of the "reietto" [outcast] and structurally through the articulation in letters of the protagonist's feelings. Love and death are the two main narrative themes featured in the novel, and in this Verga followed the conventions of the epistolary genre, although in Verga's amorous discourse the letter's enhancement of the other person's absence is, more than "a metonym of the beloved's body," a re-enactment, sad and heart-wrenching, of the protagonist's tragic death and rejection by her family, the true destination of her short-lived sentimental journey (Kauffman 20). Familial love rather than Eros connotes the sentimental discourse of the novel. Her beloved, Nino, incompletely represents the love Maria aspires to but is denied. Her letters are not addressed to Nino, and she does not transgress the social rules imposed on a woman of her age and social class. She is portrayed as an innocent victim of society, and it is her trampled innocence that bestows the narrative with a moral significance that was almost universally praised by contemporary readers. While there is a love story in *Storia di una capinera* – Nino initially woos Maria but ends up marrying her stepsister, breaking the protagonist's heart and destroying her hopes for happiness – it is on the sacredness of familial ties and society's moral obligations that the narrative drama is thematically focused. To make this point, Verga describes Maria's desire to love and be loved almost exclusively outside the convent, when, during a cholera epidemic, she lives with her family but is made to feel like a stranger. Sentimental love and family affection are closely interrelated. Verga makes explicit that a double "crime" is committed against Maria: she is deprived both of her right to be loved by her family and of her "natural" right to become a mother and a wife. "Oh! Benedetto il santo affetto di una madre che si rivela tutto in una parola e in una carezza! Benedetta la felicità che ci produce la felicità dei nostri cari!" (37) [Oh, how blessed is a mother's holy love, entirely revealed in a single word or caress. How blessed is happiness, produced by the happiness of our loved ones!] (70), Maria comments in a Cinderella-like tone while watching her stepmother behaving affectionately towards her stepsisters. Maria, tragically, will never be able to experience the joys of maternal love. To the reader of the time this must have seemed the most atrocious of fates, as, by the middle of the nineteenth century, belief in a women's natural maternal role was pervasive in both literary and journalistic publications. That the convent could never replicate the life of a family is made clear by Maria in one of her first letters to her friend Marianna:

> Mi dissero che mi davano un'altra famiglia, delle altre madri che mi avrebbero voluto bene … È vero, sì … Ma l'amore che ho io per mio padre mi fa comprendere che ben diverso da quello di queste madri che s'impongono, sarebbe stato l'affetto della madre mia propria. (6)
>
> [They said they were giving me another family, and other mothers who would love me … Yes, that's true … But the love I feel for my father makes me realize how different my poor mother's affection would have been from that given by those appointed mothers.] (27)

The idea of the sacredness of the family reverberates throughout the novel and is expressed by a set of contrasts (nature-nurture, life-death, etc.) that highlights the indispensable role played by the family in the individual's pursuit of happiness, as well as in the collective striving towards progress and social order. Convent life is associated with the "silenzio" [silence], "solitudine" [solitude], and "raccoglimento" [meditation] of its "poveri carcerati" [poor prisoners], while the family provides those pleasures "che il buon Dio ha dato a tutti: l'aria, la luce, la libertà" [that the good God had given to everyone: air, light, and freedom] (9, 7). Verga embraced the current bourgeois cultural and social ideology of the family as well as a secular scepticism towards religious institutions – a choice that, in narrative terms, translated into a positive portrayal of the family as the place where the individual could find refuge from modern social ills. Symbolic of the constraints and hollowness of modern society, the convent is contrasted, significantly, not to Maria's own family but to that of a modest man, the *castaldo* [steward], whose household appears like a mirage of happiness in her personal landscape of emotional desolation. Caught between the realities of a convent falsely disguised as her new family and her own family, who are consumed by economic concerns and unable to truly love her, Maria praises the steward's household for its unpretentious demonstration of simple but genuine family values:

> Dall'altra parte dello spianato c'è una bella capannuccia col tetto di paglia e di giunchi, ove abita la famigliola del castaldo. Se vedessi la bella capanna, com'è piccina ma pulita! [...] Mi pare che codesta famigliuola, riunita in due passi di terreno, debba amarsi dippiù ed essere maggiormente felice; mi pare che tutte quelle affezioni, circoscritte fra quelle strette pareti, debbano essere più intime, più complete; che il cuore commosso e quasi sbarlordito dal cotidiano spettacolo di codesto orizzonte, ch'è grande, trovi un gaudio, un conforto nel ripiegarsi in se stesso, nel rinchiudersi fra le sue affezioni, nel circoscriversi in un piccolo spazio fra i piccoli oggetti che formano la parte più intima di se stesso, e che debba sentirsi più completo trovandosi più vicino ad essi. (8)
>
> [On the far side of the lawn there's a pretty cottage with a roof of straw and rushes, where the steward's little family lives. If only you could see it – you'd see how tiny it is, and yet so clean, and how neat and tidy everything is there! ... I think that family, living together on those few square feet of land, must love each other all the more and be much happier; that in that limited space all their feelings must be deeper, and more absolute; that to a heart overwhelmed and almost bewildered by the daily spectacle of that vast horizon, it must be a joy and a comfort to withdraw in itself, to take refuge in its affections, within the confines of a small space, among the few objects that form the most intimate part of its identity, and it must feel more complete in being near to them.] (Verga, *Sparrow* 28–9)

The description of the steward's family is highly romanticized and, seen from the point of view of Maria, who looks at it from her "scatolino" [little box] – or casket (symbolic of her imminent death) – it evokes a strong feeling of nostalgia for her youth and for the natural condition of happiness provided by the family security she once enjoyed when both her parents were alive.

Not knowing whom to turn to, Maria confides in a fellow student at the convent, Marianna, to whom she writes during a brief stay at her family's country estate, and when she returns to the convent after realizing that her friend has gone back to live at home. Confidantes are important agents in epistolary communication, and Marianna plays here the role of the passive confidante; absent by definition, she cannot witness the narrated events and has to be told about them; like all confidantes she "is most fundamentally an archivist" (Altman, *Epistolarity* 53). She silently receives and files epistles (to which she never replies – but her silence eloquently defines the sense of her presence that the letters seek to create) in an imaginary repository of knowledge. Though almost invisible, Marianna is an essential link in the narrative, because it is through her that readers gain access to Maria's inner sorrows. Through this archive of information, the external readers are made to witness the events that unfold in the epistolary narrative. The epistolary composition of the novel, thus, though enhancing the intimate tone with which Maria delivers her sorrowful confidential messages, has strategically here a much broader role than that of allowing the protagonist to "unburden her heart," confessing to a silent though sympathetic listener the pain that afflicts her body and soul. Maria needs an active listener to give meaning to an otherwise senseless story of familial abandonment. Although both letter and diary, as narrative strategies, could have provided Verga's novel with a similar atmosphere of intimacy and confession, yet it was the letter that effectively enabled the author to avoid an introspective monologue and to elicit the readers' empathy. Maria's letters invite readers to enter empathically and directly into the fictional world described in the novel. Empathy, as the philosopher Laura Boella reminds us, derives from the Greek *pathein* (which means to suffer, to be in pain) and should not be confused with sympathy or compassion: "l'empatia è l'atto attraverso cui ci rendiamo conto che un altro, un'altra, è soggetto di esperienza come lo siamo noi, vive sentimenti ed emozioni, compie atti volitivi e cognitivi" [empathy is the act through which we realize that someone else is the subject of an experience, just like us, who feels sentiments and emotions, and experiences willing and cognitive acts] (xii). While compassion implies a unilateral acknowledgment of someone's experience, empathy makes the other's experience a part of our own. It is an act that permits one to recognize the existence of the other and his or her experience; it is also a feeling that creates awareness of the other through an emotional involvement, "nella condivisione, nel superamento della distanza" [in sharing with the other, and overcoming distance] (Boella xvi). From a structural point of view, the letter facilitates the emotional involvement of the reader and also both enunciates and closes the distance between interlocutors (a symbolic measurement of the protagonist's suffering), allowing the (real) reader – and the writer, as well – to enter and at the same time stay outside the narrative world. Both writer and reader stand outside the narrative but have knowledge of the events narrated in the letters. In this sense, the letter as a literary strategy may be understood as Verga's early experiment with literary genres in order to create a narrative in which the author is a dispassionate observer and the reader an involved witness, someone who can make a personal evaluation of the moral and artistic proposition on the basis of his or her experience with or knowledge of the matter presented in the letter. In the introduction, Verga announces what he is about to present: "una di quelle intime storie, che passano tutti i giorni inosservate" [one of those intimate stories that pass unnoticed every day], thus bringing to our attention the reality of Maria's inner suffering that, he thought, needed not only to be told but also listened to, exposed to the indifference of modern society (Verga, *Storia di una capinera* 3; *Sparrow* 23). As he was in Florence, thinking about his family and writing to his mother to ask about certain family memories of convent life, Verga created a correspondence that he drew on for both content and style. The language of a familiar written exchange, the family stories about convent life, and the intimacy of the memory of family life evoked nostalgically from far away are all elements that belonged to Verga's experience of being away from home and that he wove into the story of Maria. Written in the epistolary form and thematically developed as the tragic story of a young woman cut off physically and emotionally from her family, *Storia di una capinera* is a novel for which Verga relied on his own experience of nostalgia and emotional displacement, which he frequently expressed in his letters to his family, to produce a narrative of denunciation of a phenomenon – the isolation of the individual in modern society – that he began to observe during his very first stay in Florence and that he was eager to "gettare in faccia" [throw in the face] of his readers.[12]

The topic of the convent, as I discuss later, certainly contributed to the emotional effect of the story, but Verga's formal choices, his familiarity with letter-writing conventions (the use of certain linguistic and stylistic forms), made the letter an apt choice of literary device. As Maura Brusadin suggested in the editor's introduction to the novel, Verga's linguistic and thematic choices are inextricably intertwined ("lingua e struttura tematica procedono in Verga in funzione reciproca") (xxxv). If, on the one hand, Verga used themes such as destiny, family, and the role of nature in human existence that he later developed more fully in his *verista* productions, on the other hand, the variants that appear in different editions of *Storia di una capinera* demonstrate his attention to, even obsession with, the formal aspects of his novel (written in a language that was modelled after spoken Florentine and that he continued to correct and change) and his determination to create a text that stylistically reflected his thematic choices. "Ascoltando, ascoltando si impara a scrivere," Verga remarked. "Lo stile non esiste fuor dell'idea. Se lo stile consiste massimamente nella forma del periodo, esso deve adattarsi all'idea, deve vestirla, investirla" [Listening, listening, that is how one learns to write. Style does not exist outside of the idea. If style consists mainly in form of the period, it must adapt to the idea, it must clothe it and fit it] (Ojetti 65). Verga infused his novel with a language and style that, although at times highly melodramatic, reflected a standard Italian conversational register that, as was recommended in epistolary manuals of the day, sought to replicate natural speech without disregarding the stylistic norms appropriate for written expression. "Ognuno dunque scrivendo lettere si ricordi in primo luogo che mette in carta un discorso" [Everyone therefore while writing letters must remember first of all that they are presenting a speech on paper], as Mestica reminded his students of epistolary writing, and in a similar vein, Verga used the letter form in his *Capinera* to give his writing the spontaneity of speech (4). As Maura Brusadin noted, in the original version of Verga's novel (the one that appeared in the periodical), Maria's confidences "sono spesso affidati a frasi e uscite tipiche del parlato" [are often expressed in phrases and forms typical of the spoken language], while in the published book, the oral aspect of the language is mediated by the use of more polished expressions "nel tentativo di creare un linguaggio omogeneo ed 'aggraziato' che convenga alla forma scritta" [in the attempt to create a homogeneous and graceful language that is suited to the written form] (xxiii). Literary as Verga tried to make his novel, the oral, conversational aspect of his writing is what he mainly focused on in his artistic search for authenticity, and that was what readers seemed to appreciate most, recognizing in it perhaps the real originality that gave the novel its unexpected linguistic and stylistic modernity. The novel, written in the form of letters and in a language with which readers identified, resonated as an authentic reflection of the cultural realities of the time. "L'argomento non è nuovo," wrote Giorgio Baseggio in *Perseveranza* on 14 September 1873; "Ma è nuova la forma, difficilissima, seguita dall'autore. Egli ha prescelto l'autobiografia epistolare" [The topic is not new, but new and very difficult is the form adopted by the author. He chose epistolary autobiography].[13] While the epistolary genre was not a novelty to Italian readers, there was a perception among contemporaries that Verga used the letter form to infuse his novel with a "narrazione semplice e affettuosa, tutta spontaneità, tutta verità" [a simple and affectionate narrative filled with spontaneity and truth].[14] Authenticity and naturalness had traditionally been essential attributes of epistolary writing, but Verga made his Maria sound like a spontaneous letter writer taken out of the readers' world. "Letters," as Elizabeth Goldsmith pointed out, "were increasingly valued for their 'natural,' 'authentic' and purportedly inimitable qualities, and good letter writers were said to be those who could make their letters 'seem to speak' in a plain and unpedantic style" (47). Verga was aware of these pre-existing literary conventions but adapted them to his artistic ambition of creating a modern literary voice; to be a good "letter writer" meant to endow his epistolary text with specific linguistic and stylistic choices that made his narrative text sound like a real letter, like a letter readers themselves might receive. As one reviewer, D'Ormerville in *Il Pungolo*, commented, "Lo stile è facile, la lingua italianissima, ma di quella che tutti conoscono, di quella che si parla dalle persone ammodo, non di quella che si pesca a fatica nei vortici della Crusca" [The style is simple, the language is certainly Italian, but one that everybody knows, one that is commonly spoken in society, not one that people can hardly catch from the vortexes of the Crusca], and it was not uncommon for reviewers to admit that they had read the novel "tutta di un fiato" [straight through].[15] Readers liked his novel because it resonated with the world they lived in, and they could identify with the language, ideas, and characters that emerged from it. Verga's experiment with a literary language that came across as "viva, compita, spontanea" [alive, polished, spontaneous] – attributes that fifteen years earlier Bonghi had longed to see in a modern writer – had the effect of producing, perhaps for the first time, a real reading public, the concrete embodiment of an otherwise idealized readership coveted by Italian writers throughout the nineteenth century. The emergence of this public of readers may also be conceptualized in terms of the emotional effect of the novel – an effect that was comparable, according to the English writer D.H. Lawrence, to that of the sentimentality of other European writers, such as Dickens and Hardy (Ghidetti 189). If the letter as a literary strategy enhanced an empathetic reaction among readers, the epistolary novel as a whole also fostered a sense of shared identity, based on a consensus of emotional understanding. As Suzanne Keen points out, "There is no question that readers feel empathy with (and sympathy for) fictional characters, and other aspects of fictional worlds [...] readers' and author's empathy certainly contributes to the emotional resonance of fiction, its success in the marketplace, and its character-improving reputation" (vii). Thanks to the inherent property of correspondence whereby the "I" always addresses, implicitly or explicitly, the "you," the letter format turns the narrative expression of a personal story into a shared experience, a social and cultural encounter. The publication of *Storia di una capinera* may be considered one such social occasion, a cultural event that stimulated the circulation of ideas and a new awareness of cultural identities.

Famously, Verga insisted that his work was more "intimate" than "social": "Sto lavorando ad un romanzo intimo di cui sono contento sino a quest'ora" [I am working on an intimate novel which so far I am happy about], he wrote to his mother on 26 June 1869 (*Lettere sparse, 28*), and much scholarship has been produced endorsing the author's intimate poetic approach. If we look carefully at the text, however, we notice that these two elements are not so neatly distinguished. In his introduction – a manifesto for his narrative poetics – he wrote,

> la storia di una infelice di cui le mura del chiostro avevano imprigionato il corpo, la *superstizione* e l'amore l'avevano *torturato* lo spirito; una di quelle *intime* storie, che passano *inosservate*, storia di un cuore tenero, timido, che aveva amato e pianto e pregato [...] ed io pensai alla povera capinera che guardava il cielo attraverso le grate della sua prigione, che non cantava, che beccava tristemente il suo miglio, che aveva piegato la testolina sotto l'ala ed era morta. Ecco perché l'avevo intitolata: *Storia di una capinera.* (3-4; emphasis mine)
>
> [the story of an unhappy girl whose body had been imprisoned within the walls of a convent, and whose spirit had been tortured by superstition and love – one of those intimate stories that pass unnoticed every day – the story of a shy and tender heart, of one who had loved, and wept and prayed ... And I thought then of the poor songbird that would gaze at the sky through the bars of its prison, that would not sing, that would peck sadly at its grain, that had tucked its head under its wing and died. That is why I titled it *Storia di una capinera* – The story of a Songbird.] (*Sparrow* 23)

Words like "superstition," "tortured," and "unobserved" place the emphasis on the suffering and the social injustice perpetrated against this young girl, sent against her will to a cloistered life, while "tender heart" and "intimate story" are presented as the poetic inspiration from which sprang the artistic creation. And yet these two narrative elements are not as antithetical as they might seem at first glance, especially if one looks at the social aspect in terms of an enunciation rather than a denunciation of reality. In other words, the social aspect is used here by Verga to contextualize the emotional import of his narrative, to enhance empathy among readers by creating a sense of shared feelings and a vivid awareness of what he perceived as an emerging social reality in which the individual was being trampled by the negative forces of modern society that privileged, even within the realm of the family, the rationale of profit and economic advancement. "Voi siete la vittima della vostra posizione, della cattiveria della vostra matrigna, della debolezza di vostro padre, del destino!" [You are the victim of your position, of your evil stepmother, of your father's weakness, of destiny!] (40) Nino tells Maria after it becomes clear that she is inevitably destined for the convent. And Maria echoes Nino's words of condemnation by lamenting with Marianna the injustice done to her: "Quanto siamo meschini, amica mia, se non possiamo essere giudici della nostra istessa felicità!" [How miserable we are, my friend, if we cannot be the judges of our own happiness!] (42). A sequence of exclamation points punctuates the dramatic crescendo, infusing the narrative with a sentimental tone that overcomes at times the note of authenticity the author aimed at, but one that nevertheless resonated with contemporary readers. Thematically based on the sorrows of a young woman forced to become a nun, *Storia di una capinera* became a commercial success thanks to the timing of its appearance on the cultural market. The theme was topical and well known to Italian readers, not only through Denis Diderot's *La religieuse* (1760), Tommaso Grossi's *Ildegonda* (1820), Giovanni Rosini's *La monaca di Monza* (1829), and, naturally, Alessandro Manzoni's *I promessi sposi* (1840), but also through newspaper articles. It was in the pages of these newspapers, where parliamentary debates over the suppression of religious orders and institutions were discussed, that postal culture thrived (by way of reviews, fiction, or articles in letter form) and readers shared their interpretations and appreciation of Verga's version of an unwilling nun's story.

Two texts in particular, Enrichetta Caracciolo's *Misteri del chiostro napoletano* and Caterina Percoto's "Memorie di convento," both published in 1864, had galvanized public opinion over the abolition of religious institutions and were most likely known to Verga through his contacts with intellectuals living in Florence and Milan and the reviews appearing in periodicals. As one anonymous reviewer of Caracciolo's best-selling book – signed with R.B., and identified by Alfonso

Scirocco as Ruggero Bonghi – put it, Manzoni's portrayal of the "monaca di Monza" had appeared as something from a distant world to contemporary readers, while Caracciolo's description of her life in the convent resonated as something that could happen to them or to someone they knew. In this sense, Caracciolo's memoir anticipated some of the crucial elements that later readers would find in *Capinera*, among them the authenticity of the representation.

According to R.B., it would have been difficult for readers to identify with Manzoni's story of a nunnery because it was a

> storia antica, tratta di monache e di monasteri che il tempo e la civiltà spazzarono via; questa [Caracciolo] è storia contemporanea che si vede e si ripete ogni giorno intorno a noi; tormentatori e tormentati sono vivi, accusati e accusatori stanno in faccia gli uni agli altri, aspettando un giudizio che i legislatori della nazione stanno per essere chiamati a pronunciare. (Scirocco 230)

> [ancient history, dealing with nuns and monasteries, which time and civilization have swept away; this is contemporary history, which repeats itself every day as we can see all around ourselves: tormentors and tormented are alive, accused and accusers face each other, waiting for the nation's legislators to pronounce a judgment to which they are being called.]

The contemporary context rather than the topic itself was the crucial element in Caracciolo's success, as it would be for Verga only few years later; and if, as Federico De Roberto complained, contemporary Italian literature lacked thematic variety to the point that one found in modern novels "presso a poco la stessa cucina, lo stesso tema, e se ne annoja" [more or less the same recipe, the same theme, and gets bored with it], the readers of Caracciolo's autobiography didn't seem to mind the old "nunnery" story and, on the contrary, followed with great interest the tormented vicissitudes of someone they could identify with – a woman who, forced by circumstances to take the veil, made the public discussion on convents, by her own testimonial, all the more relevant (Ojetti 84). As Italy was going through social reforms, public opinion was galvanized by topics of social modernization, and publications like this one greatly resonated in a country uncertain about what course to take, especially in matters dealing with women's education and the role of religious institutions in modern society.

With the unification of Italy in 1861, the Italian government extended to the national territory a law that, passed initially in Piedmont in 1855, suppressed several ecclesiastical institutions in order to relieve the state of its financial commitments to religious corporations. In 1864, parliamentary discussions on the relationship between state and church continued, as many politicians considered the expropriations of ecclesiastical properties at the national level to be a lucrative opportunity for a new nation in dire need of funds. On 18 January 1864, Giuseppe Pisanelli, minister of "Grazia, Giustizia e Culti" [Grace, Justice and Religion], proposed legislation "per la soppressione delle corporazioni religiose" [for the suppression of religious institutions], and later that year, during the fall session, the new government, headed by Lamarmora, presented a new legislative proposal that became law in 1867.[16]

In the time between the two proposed laws of 1864, Enrichetta Caracciolo's *Misteri* was published by Gaspero Barbèra in Florence. Its commercial success was enormous, and it became "il caso letterario del momento" [a literary cause célèbre] (Scirocco 220). Barbèra recounted in his memoir the international sensation produced by this publication:

> Questo libro destò universalmente in Italia, in Francia, in Inghilterra, in Germania e in altre parti (credo persino in Russia) un vero e grande entusiasmo. Quante edizioni abbia io fatte non saprei dire; posso assicurare che sono state molte, e che le domande di tradurre questo libro nelle varie lingue d'Europa furono diverse. (280)
>
> [This book sparked great enthusiasm universally in Italy, France, England, Germany, and other places (I believe even in Russia). How many editions I have made I cannot say; I can only assure you that there were many, and that the requests to translate this book into various European languages were numerous.]

In its fifth edition by the end of 1864, and translated into French, English, Spanish, German, Hungarian, and Greek, *Misteri* was the autobiography of a woman of the noble family of Caracciolo de' Principi di Forino of Naples, who, after losing her father at a young age, was forced by her mother (out of dire financial necessity) to enter a convent in order to save the family patrimony – a practice sanctioned among the nobility by the tradition of marrying only the first-born daughter. After many abuses and injustices suffered at the hands of ecclesiastical authorities, Enrichetta Caracciolo managed to leave the convent and to write her memoir, presented also as a pamphlet against the perils of this old ecclesiastical practice. The manuscript eventually reached the desk of a major publisher, Barbèra, though the editorial history of the book illustrates the many complications the author faced, even after leaving the convent, in trying to make her story known. Barbèra, in fact, had initially received the manuscript not from her but from a man ("tal cavaliere Spiridone Zambetti" [a certain Mr. Spiridone Zambetti]) who had edited the original text and omitted the author's name (280–2). Barbèra liked the book instantly and proposed its immediate publication, conditional on first getting in touch with the woman who had written it. In the end, Caracciolo agreed to Barbèra's request to have her name appear on the title page and to split the payment (of 1,400 Lire) between her and Zambetti. It is not clear whether Enrichetta Caracciolo had given Zambetti her book only to read or whether she had entrusted him with the task of looking for a publisher and asked him to omit her name so that she would not expose her family to public scrutiny and likely criticism. In the end, however, Barbèra insisted that for an autobiography to be successful it had to convince readers of the authenticity of the experience and the reliability of the speaker, and persuaded her to have her name appear on the book cover. And he was right. The fact that the story was related by a member of the Caracciolo family, and that not even a woman from the nobility could defend herself against such tyranny, made the story all the more compelling to readers, not only in Italy but in many other countries.

In addition to the prestige of the author's family name, the picaresque tone of the narrative – with the heroine, against all odds, escaping from great peril – certainly contributed to the book's success. The account of the brutalities inflicted on women by religious institutions, with no spiritual justification, resonated deeply with the public. Presented by the author herself as a testimony intended "confermare, quanto è da me, con argomenti di fatto, l'opportunità e la giustizia del Decreto col quale si sopprimono dal Governo italiano i conventi, e disingannare a un tempo coloro, se pur ne restano ancora di buona fede, che tenesser quei luoghi per asili di tutte le religiose virtù" [to confirm, with what I myself can say and actual facts, the opportunity and justice of the law with which the Italian government is going to suppress convents, and at the same time open the eyes of those, if there are any, who still believe in the good faith and religious virtue of those places] (viii), the book sparked a lively debate among supporters and critics of the proposed law. Several newspapers, including *La Nazione*, *Opinione*, and *Il Pungolo*, reviewed the book soon after it came out, focusing on the "significato politico-morale della monaca" [political and moral meaning of a nun] who, after decades of forced confinement and persecution, was finally able to escape the torments of convent life in 1854 and leave the Benedictine order and who, upon Garibaldi's arrival in Naples on 7 September 1860, cast off her veil and, in a symbolic gesture, returned it to the altar at the Duomo in Naples (Scirocco 222). "Non è esprimibile la profonda espressione che ha fatto in Italia" [One cannot express the profound impression that it made in Italy], wrote a reviewer in *L'Italia* on 24 August 1864, describing the impact that the book made on the public perception of what was considered mainly a moral rather than a religious question. Ferdinando Martini also insisted on the moral and political significance of the book: "Il libro della signora Caracciolo mostra come e quanto efficacemente possano essere rivoluzionarie le donne, quanta parte esse possano avere nella lotta della verità contro l'assurdo" [Mrs Caracciolo's book demonstrates how effective and how revolutionary women can be, what a significant role they can play in the battle for truth and against the absurd] (*La Nazione*, 22 August 1864, 2). In a long review of Caracciolo's book, published in *La Nazione* on 20 and 22 August 1864, Martini emphasized the innately unnatural conditions in which monks and nuns were asked to live. Monks lived in idleness and apathy, he wrote, and women – in accordance with the emerging nineteenth-century ideology of female domesticity – were deprived of their most natural mission in life: living in and caring for the family. "Che fa della donna la legge monastica?" [What does the monastic law do to women?], asked Martini. "Essa la prende fanciulla, la separa dal mondo, le ordina di lasciare il suo nome, ultimo vincolo che la stringa ancora alla sua famiglia, e chiusala in uno di quei sepolcreti di vivi, che si chiamano conventi, la toglie dalla sua azione di moglie e di madre, per la quale essa è nata" [It takes the young girl, separates her from her family, orders her to abandon her name – the last connection to her family – and after closing her in one of those tombs for the living, which they call convents, takes her away from her mission as wife and mother, for which she was born] (*La Nazione*, 20 August 1864, 2). Doubly heroic, first for escaping tyranny and also for having had the courage to denounce it in public, Enrichetta Caracciolo was to be commended, Martini suggested, for advancing progress in society with her inspirational work of literature, which, he added, if it did not excel in artistic originality certainly had the merit of sparking a lively debate among readers and increasing their awareness of the country's ills.

> Siano le benvenute adunque le Memorie di Enrichetta Caracciolo! Siano le benvenute e diano un altro crollo alle mura de' chiostri, scaglino un'altra

> pietra contro l'idolo dell'errore, e soprattutto vadano per popolo, perocché là dove il popolo non è, non si fa veramente opera grande e durevole. (*La Nazione*, 22 August 1864, 2)
>
> [May Enrichetta Caracciolo's memoir be welcomed, then! May it be welcomed and may it give another push to the convent's walls, and may it throw another stone against the idol of error, and above all may it be welcomed by the people, because without the people a work of literature will never be great and enduring.]

It is this moral judgment that ought to be kept in mind for Verga's novel as well, as readers in the early 1870s might very well have remembered the outrage provoked by Caracciolo's memoir, and have made a similar moral rejection of a social practice that stifled individual aspirations for fulfilment. In other words, readers approached Maria's story with the same sense of outrage as they had that of Enrichetta Caracciolo, and even if by then women were no longer forced into convent life (at least not in a socially accepted way), the issues of female education and marriage were anything but resolved.

While Enrichetta Caracciolo was galvanizing public opinion with her memoir, another woman, Caterina Percoto, had also published her personal account of convent life. A known public figure and writer, Caterina Percoto, famous for her "novelle rusticali," had been educated by nuns, as many girls of the upper classes were – an experience that had a strong impact on her life. From April to June of 1864, she published, in *Giornale delle Famiglie,* four articles written in letter form titled "Memorie di convento."[17] Presented as a private conversation between two friends, Percoto's letters on convent education adopted the intimate tone of a personal confidence: "Le cose che io ti dirò, le cavo dal cuore, e dalla triste esperienza dell'educazione che io stessa ebbi la disgrazia di subire" [The things I am about to say come from my heart, and from the sad educational experience I had the misfortune to be subjected to].[18] Recounting "i miei brutti sette anni,"[my ugly seven years] spent in a convent near Udine from the age of ten, in "un'edificio severo, lontano dal centro della città, isolato con rade finestre sempre cieche e riparate da diversi ordini di grate" [a severe and isolated building far from the centre of the city, with few, obstructed windows, protected by several layers of bars], Percoto set out to illustrate the negative effects of convent education, focused primarily on prayers, ascetic meditation, and practical domestic skills such as sewing (78). Little or no formal education was provided by the nuns, who were, with a few exceptions, largely uneducated themselves. "Non era loro colpa" [It was not their fault], Percoto rushed to explain, if the nuns did not truly feel the "missione di educare e d'istruire. Vi si erano sobbarcate per necessità non per vocazione" [mission to educate and instruct. They went into it out of necessity and not as a vocation].[19] Because in the northern regions of Italy some religious orders had already been suppressed, the nuns who wanted to continue to live in the convent had to change the mission of their order from religious to educational, from that of an "istituto religioso" [religious institute] to that of a "collegio per l'educazione della fanciulle del quale nella provincia c'era assoluta mancanza" [college for the education of girls, which in the province was absolutely needed].[20] Remembered as one of the most painful periods of her life, along with the death of her father, Percoto's convent education was not presented in completely negative terms, however:

> Monache e compagne, misteri di quel recinto inesplorato, gioie e pensieri di una solitudine inaccessibile quante volte nella fantastica vostra forma e con tutti i vostri bizzarri colori non mi passaste voi per la memoria? Li metto qui sulla carta e saranno una biografia di quell'epoca cara insieme e dolorosa al mio cuore![21]
>
> [Nuns and fellow students, mysteries of that unexplored place, joys and thoughts of an inaccessible solitude, how many times have I remembered you in all your fantastic forms and bizarre colours? I put them here on paper and they will be a biography of an era, at once dear and painful to my heart!]

The tone of her criticism, unlike Caracciolo's, was never accusatory or denigrating towards the institution. Percoto viewed her letters more as part of an autobiography she had been meaning to write since the 1850s than as a criticism of religious institutions or a polemical pamphlet in support of the legislation being discussed in the parliament. She, therefore, vehemently argued with Lampugnani when the publisher, complaining that her letters were written in a style more appropriate for a book than a newspaper article, tried to edit them and to add some "drama" in order to make them more appealing to readers. Her memoir of convent life was supposed to continue beyond the fourth letter, but Percoto decided to suspend publication. As she declared in a letter to a friend and fellow writer, Luigia Codemo,

Figuratevi precetti a me che non sono mai stata a scuola e che ho sempre scritto e voglio scrivere proprio e precisamente a mio modo. Se le cose mie non gli piacciono non le pigli, ma non s'imponga di farle pensando il gusto di nessuno [...] Dove il Lampugnani ha ragione di lagnarsi di me è per l'interruzione di quelle lettere. Ma non mi fu possibile evitare in che la politica e gli eventi che precipitarono mi persuasero a sospendere il lavoro che diventava inopportuno così per me come per il sig. Lampugnani e credo anche pel pubblico.[22]

[Imagine that he wanted to instruct me, who has never been schooled and have always written, and always want to write, precisely in my own way. If he does not like my things he should not take them, but he should not ask me to write according to the taste of anybody else. Lampugnani is right in complaining about one thing only, that I stopped writing the letters. But I could not avoid it because the politics and the events that followed persuaded me to suspend my work, which was becoming inappropriate for both myself, Mr. Lampugnani, and, I believe, for the public.]

The fact that, in the middle of the 1860s, convents were a major topic of public discussion does not in itself establish that Verga intended *Storia di una capinera* to add fuel to the social debate. He did not want his novel to appear as a social pamphlet, but that does not mean that he was not aware of or influenced by the contemporary debate. In addition, there is the question not only of how he intended the work to be perceived but also of how it was in fact perceived by readers. Dall'Ongaro's introduction to the first edition of *Storia di una capinera* certainly made a connection to the social aspect of this theme; in fact, we could even speculate that his enthusiasm for Verga's novel was triggered by that type of reading. And perhaps because the novel was dedicated to Percoto, a writer known not only for her articles in *La Ricamatrice* but also for her short stories, several of which dealt with convent-related themes, Dall'Ongaro explicitly underlined the novel's topicality. Furthermore, Percoto's own letter to Verga, written soon after the publication of the volume, suggested that, on account of his novel, Verga was perceived as a "champion" of a social cause.

Ella ch'è giovane e ch'ebbe in dono dal cielo una parola così simpatica, così vera e così efficace, si faccia nostro campione. L'Italia gliene sarà riconoscente, e io e il Dall'Ongaro saremo ben lieti d'essere stati fra i primi ad estimarla come uno dei nostri più valenti scrittori. (De Roberto, *Casa Verga* 168)

> [You – a young writer who was given by heaven the gift of words so congenial, so real and efficacious – be our champion. Italy will be grateful to you for it, and Dall'Ongaro and I will be happy to have been among the first to recognize you as one of our most talented writers.]

Readers identified with Verga's novel because of its theme but also because of its writing style: authentic, spontaneous, real. Its authenticity and its subject matter seemed to be the two main reasons for the novel's popularity with readers. It would therefore be almost naïve to discount the "social" origin of the novel's success (a success that Verga was quite happy to enjoy), given that most of Verga's contemporary readers did connect the novel to the moral problems associated with religious institutions. Sergio Pautasso may be right in suggesting that *Storia di una capinera* "è diventato un romanzo con caratteristiche diverse da quelle che avrebbe voluto che trasparissero dalla vicenda narrata" [became a different novel with different characteristics from what he may have intended] (ix). But Verga must have been aware of these characteristics all along, and was not willing to renounce the benefits (in terms of audience appeal) that came with them. When he publicly denied the social aspect of his novel, he was only trying to protect it from being viewed as a political pamphlet or a work of literature with little artistic merit.

The overwhelming majority of contemporary reviews saw Verga's novel as a literary work infused with social and historical relevance. Dall'Ongaro, his first reader and supporter, described *Storia di una capinera* in a letter to Treves in 1869 as an "argomento di attualità palpitante e studio fisiologico e patologico di un cuore che si spezza" [a topic of great currency, a physiological and pathological study of a broken heart], concluding that "Il Verga sarà, credo, il migliore dei nostri romanzieri sociali" [Verga will be, I think, the best of our social novelists] (quoted in Gambacorti 173). Dall'Ongaro was not alone in giving Verga's novel a social and moral interpretation. For instance, Angelo De Gubernatis wrote in *Rivista Europea*, "Quanta verità e potenza di sentimento! Quanta progressiva corrispondenza fra gli affetti e le parole! E tutto il libro può considerarsi come la più eloquente orazione funebre per la soppressione dei conventi" [How much truth and powerful sentiment! How much progressive correspondence of effects and words! The whole book may be considered a most eloquent death oration for the suppression of convents] (1 September 1873); while B.E. Maineri in the pages of *La Lombardia* invited readers to recognize in the novel the overcoming of entrenched superstitions: "leggete pure le lettere di suor Maria, ché la storia di lei, storia intima, è di quelle che si potrebbero contare a migliaia, poiché le vittime della superstizione sono forse più numerose delle arene del mare [...] Questo libro è un'elegia; leggendolo, leggerete di pianto soave per una vittima della tirannide domestica e della superstizione monastica" [read sister Maria's letters, because her story, an intimate one, is one of those which could be counted in the thousands, because the victims of superstition are perhaps more numerous than sand at the beach ... This book is an elegy; reading it, you will shed sweet tears for a victim of domestic tyranny and monastic superstition] (3 November 1873). Pacifico Valussi, moreover, in *L'Universo Illustrato: Giornale per Tutti,* urged parents to read the novel, stressing the realistic and educational value of the narrative:

> È un libro da raccomandarsi soprattutto ai genitori. All'autore diciamo che egli ha già contratto degli obblighi verso l'Italia [...] Il racconto che ora tende a diffondersi nei giornali e per i giornali deve avere anche quest'ufficio di far conoscere meglio l'Italia e gl'italiani a loro medesimi, di attutire le passioni politiche [...] di condurre i nostri compatrioti alla riforma e rigenerazione morale della società mediante l'arte ispirata dalla natura, mediante la pittura che ritrae dal vero meglio che dalla sbrigliata fantasia che spazia i campi dell'immaginario [...] Faccia il Verga come fece prima di lui Caterina Percoto che interessò l'Italia intera e si fece leggere e tradurre anche dagli stranieri, descrivendo la semplice vita e gli affetti dei contadini friulani. Ci racconti e descriva l'isola sua, e i suoi racconti torneranno graditi, fino in questa regione subalpina ch'è confine all'Italia e dalla quale gli mandiamo come ad un ignoto amico un cordiale saluto. (28 September 1873)
>
> [It is a book to recommend especially to parents. To the author we say that he has already discharged his duties to Italy ... The story which tends to circulate in and for newspapers must also have the task of making Italians themselves know their Italy better, to tone down political passions ... to push our fellow citizens to reform and moral regeneration of society through an art that is inspired by nature, through a picture which portrays reality better than a fantasy unravelled from the fields of the imagination ... Verga should do what Percoto did before him, describing the simple life and affections of the peasants of Friuli. He should tell us about his island, and his stories will be welcome all the way up to this subalpine region which is at the border of Italy and from which we send him, as if he were an unknown friend, our cordial greetings.]

In his reference to Caterina Percoto, Valussi emphasized what he considered Verga's "regional" rendering of a story with national appeal, suggesting also that Verga's novel could be interpreted as a sign of renewal in the national literary tradition. "Noi attendiamo," Pacifico wrote in the same article, "speriamo dalla vita nuova dell'Italia una nuova letteratura, una letteratura più nostra, più viva, più immedesimata colla vita reale d'una società che è finalmente uscita dal letargo morboso, nel quale era stata per molto tempo con arte malvagia mantenuta. E questo libro d'un giovane è uno degli indizi della nuova letteratura italiana" [We are waiting, and hoping, from Italy's new life, to have a new literature, that is more ours, more alive, closer to the real life of a society that has finally escaped the morbid lethargy in which it was kept for a long time with evil art. And this book by a young author is one indication of a new Italian literature]. The eagerness to find new modalities of literary expression, modern and national in nature, might very well have coloured the opinion of these commentators; however, what is most striking in these reviews is the almost uniform appreciation of Verga's literary voice, of the style and language with which he had portrayed a piece of Italy's reality that people recognized as being truthful. Finally, Salvatore Farina, in *Rivista Minima,* commented that "il Verga appartiene alla scuola degli osservatori attenti, vale a dire alla scuola che non ha altri maestri tranne la natura ed il cuore" [Verga belongs to the school of careful observers, that is, to a school that has no other teachers except for nature and heart] (17 August 1873). Acclaimed by both ordinary readers and professional critics, Verga's novel resonated with an audience eager to embrace a work of literature that promised to renew the national literary tradition.

All Verga's novels written during the late 1860s and early 1870s centred thematically on love. *Una peccatrice* (1865), which he later repudiated, is the dramatic story of Narcisa Valderi, wife of the Count of Prato, who falls passionately in love with a young Sicilian artist and dies for it; *Storia di una capinera* (1871) narrates the sentimental soul searching and ultimate defeat of Maria; *Eva* (1873) is the story of a beautiful dancer who abandons her stage life and the glitter and comforts of the theatre in order to follow a young, penniless artist, with whom she falls madly in love, but who soon, owing to the monotony and financial constraints of their daily life, becomes disenchanted with her when she no longer resembles the beautiful, unreachable creature he used to admire on stage; *Tigre reale* (1873) is the story of an impossible love between a young Italian man living in Florence and a married Russian woman, who embodies, in her aristocratic sophistication, the character of the femme fatale doomed to be sacrificed on the altar of social convention; and finally *Eros* (1875), which concluded Verga's *pre-verista* phase, is a novel focused on the romantic adventures of the Marquis Alberti, who commits suicide after realizing that his debauched behaviour has caused his wife's death. Literature, for Verga "il più sacro lavoro dell'uomo" [the most sacred work for man], offered a unique lens for examining the transformations of Italian society, and "love" proved to be a thematically appropriate instrument with which to explore the conscience of his characters and also appeal to the consciousness of a nation that considered the literary tradition to be at the core of its national identity (Raya, *Carteggio Verga-Capuana* 22).

Of course, love has been a prime inspirational force for artistic creation since time immemorial, but in post-unification Italy it gained renewed power as a result of scientific, legal, hygienic, social, and cultural debates about the role of women in society, gender relations, and rapidly changing moral and social values. With the rise of the bourgeois family and the rhetoric, in particular, of maternal love, women's sexuality, morality, rights, and education came under closer scrutiny, as it was believed, in an ideological shift from the public to the private sphere, that the progress of the country depended on the creation of an orderly and well-functioning domestic realm – the family – which was seen as a microcosm of the whole society. Studies of the rise of the bourgeois family in nineteenth-century Italy amply testify to the centrality of the family as a rhetorical trope in the burgeoning debates on the cultural and social development of Italy. Mario Alberto Banti has famously identified in the Risorgimento period the development and establishment of an ideal of the family network that inspired many patriots and intellectuals in their formulation of rhetorical and literary tropes endorsing the unification of Italy. Ilaria Porciani, too, points out how, since the end of the eighteenth century, and in particular with Rousseau's valorization of the family as the most ancient and natural form of social organization and as the only antidote to the corruption of the aristocracy, a public discourse also began to take shape in Italy about the centrality of the nuclear family, especially its importance to the project of nation-building. This discourse sought to promote at the national level, culturally and institutionally, a consensus on what constituted proper social and civic behaviour for the future citizens of a modern Italy. As Porciani suggests, "Inizia allora anche una pratica discorsiva che organizza, sotto l'egida del discorso della famiglia, nuovi modi di percepire il sé e gli altri in termini non soltanto di genere, struttura familiare, classe sociale e spazi pubblici, ma anche in termini di identità nazionale" [A discursive practice began then that organized, under the aegis of the family, new modes of perception of the self and others not only in terms of gender, family structure, social class and public spaces, but also in terms of national identity] (17). Verga's sentimental narratives elaborate on the affective bond connecting the individual to the family, and demonstrate Verga's experimental approach to questions about artistic creation in relation to both thematic and stylistic choices. Whether we look at his portrayal of love or his narrative strategies, Verga's epistolary novel reflects but also anticipates some of the ideas and concerns shared by other writers at the time, some of whom are usually not thought of in connection with *Storia di una capinera*.

The *scapigliatura milanese* has been acknowledged for having influenced Verga while he lived in Milan, although even before arriving in Milan Verga was already demonstrating familiarity, perhaps even some "consonanze ideali" [idealistic affinities], with the *scapigliati* (he had read Arrighi) (Carnazzi, 2201). If we look at Iginio Ugo Tarchetti (1839–69), the most polyvalent of the *scapigliati,* for his art developed simultaneously towards naturalism and decadence (Bulzoni and Tedeschi 10), we find some of those "consonanze," especially in the sense of a shared vision of art and love as supreme forms of human expression. Tarchetti considered art to be the purest means for the expression of love. Art, in particular literature and music, according to the *scapigliato,* allowed humans to raise themselves above the materiality of their bodily sensations of carnal desire that were in constant conflict with human aspirations towards absolute beauty. In light of such an interpretation, based on the binary opposition between the ideal and the real, spirit and body, sentiment and sensation, an uncontaminated afterlife and a corrupt society, Tarchetti turned to the figure of the artist for his poetics of love, "l'enimma eterno e insolvibile del cuore umano" [the eternal and unsolvable enigma of the human heart] (Tarchetti 28). In his trilogy *Amore nell'arte* (1869), opining that "vi sono nella nostra natura due elementi che lottano per dominarsi – l'elemento fisico e l'elemento spirituale" [there are in our nature two elements which fight to dominate one another – the physical and spiritual], and that most people fail to resolve this conflict, Tarchetti concluded that "Se l'amore debb'essere infinito e gli uomini tutti amano cose finite, gli artisti sono i soli uomini che amano" [If love must be infinite and men all love finite things, then artists are the only men who love] (30). The protagonists of the three stories included in the trilogy are all artists who experience a fatal clash between their ideals and the realities forced upon them, with a resulting disassociation from life by way of death, extreme solitude, or mental illness. A main cause of this solitude is the seemingly impossible task of finding a person who embodies their spiritual and artistic aspirations. All the women they fall in love with are either not worthy of their art or essentially undesirable, physically impaired by terrible illnesses and on the verge of death (a glimpse of his better known novel, *Fosca*, also written in 1869). But *Amore nell'arte*, as Francesca Billiani noted, "is not merely an aesthetic battle cry for 'art for art's sake'; there is a *hic et nunc* account of a widespread sense of social frustration and anxiety reflecting a more general sense of dissatisfaction which, in all likelihood, mirrors the country's problematic political and economic conditions" (486–7). Tarchetti, as Verga did at about the same time with his *Storia di una capinera*, voiced the disillusionment brought about by the political unification of Italy but also reaffirmed the relevance of literary discourse within the general Italian cultural debate. More than a mere social critique, Tarchetti's *Amore nell'arte* expressed the author's narrative challenge to the prevailing teleological, positivistic, and homogenizing cultural discourses offered by a burgeoning bourgeois rhetoric, as well as his desire to elevate the role of literature, and therefore of the writer, to a space of uncontaminated truth – indeed the highest expression of human aspirations in life.[23] In this regard, it may be said that for both Tarchetti and Verga love represented a thematic lever with which they could highlight the literary value of their work and also their status as writers and privileged observers of society. *Storia della capinera* is not only a novel thematically centred on the solitude of the individual who is forced by society to become tragically disillusioned about the most sacred of loves, that of the family; it is also a novel written in the epistolary form. Verga strategically positioned himself as a dispassionate observer of society, a writer who, as an outsider, can through his art show the world what the world itself cannot yet see. And while the protagonist here is a woman, and not a male artist, Verga was expressing similar concerns about the debilitating and corrupting effects of modern society that in his novel cause Maria's body (and symbolically the Italian body politic) to decline and eventually die. The focus of Verga's narrative experiment with artistic expression is not erotic love but familial love, which he considers an essential element in the individual's search for spiritual fulfilment in life.

Another important contemporary figure of reference was Angelo De Meis (1817–91). A scientist, author, and philosopher – belonging to the Hegelian Neapolitan school – he wrote an epistolary novel, *Dopo la laurea* [After graduation] (1868–9), that was considered by Capuana to be "il Vademecum di tutti quelli che si occupano di arte e critica" [the handbook for all those who are concerned with art and criticism] (Bigazzi 323) and that sparked a lively debate over the role of literature in modern times and the influence of positivism and scientific methodology on artistic production.[24] Verga, who was in Florence in 1869 and mingled with many of the literary figures living in the city, most likely read De Meis's novel – or was aware of it. Written as an epistolary exchange between a young man, Giorgio Fumincervello, and his older friend, Filalete Chiappanuvole, the novel elaborated a philosophical disquisition on whether the natural sciences or literature provided the most effective path of inquiry for the pursuit of knowledge. Disillusioned with the pervasive materialism of recent scientific investigation, Giorgio decides to give up a medical career (in spite of his professional successes as a doctor) and turns to poetry (his old vocation) in order to unravel the mysteries of human life. In an attempt to dissuade Giorgio from abandoning medicine, Filalete argues in response that art, like religion, is a dead field of expression, meaning that it has lost, in modern society, the meaningful role it once enjoyed in ancient times and is no longer able to speak to modern societies governed by the principle of reason and empirical evidence. Giorgio is not convinced, and De Meis makes his character state that the true essence of modern times is actually fiction – "la vera poesia è la prosa" [true poetry is prose] – because it allows writers to do what positivists and materialists cannot do, look at the imaginative interstices of reality where science cannot reach – a space *in medias res* where the external and internal experiences of life join in creating a wholesome, truly natural expression of life.

> Ma se noi ci siamo nella Natura! Giacché le idee sono noi, noi stessi, il nostro proprio e vero noi – e m'intendo le idee, e non le opinioni, che sono un noi falso e villano – . La va dunque da sé che nella Natura noi dobbiamo cercarvi noi stessi, ed è chiaro che la scienza non sarà fatta se non quando noi vi ci saremo per l'appunto ritrovati. Voi dite che nella scienza ha da essere tutto esperienza; e sta bene: a condizione però che la sia una doppia esperienza, una esterna ed una interna, e tutte e due riunite e fatte ad un punto sicché non ne formino che una sola. Ma questa a voialtri positivi, materialisti o animisti che siate, è assolutamente impossibile non solo a fare, ma nemmeno a concepire, perché avete bene la facoltà di vedere le cose l'una dopo dell'altra, ma vi manca il potere di afferrarne due in una volta sola; e siete buoni a guardarle da fuori, ma dentro – in medias res – non vi ci sapete metter mai. (11)

> [But we are in Nature! Since the ideas are us, ourselves, our own true being – and I mean the ideas, and not opinions, which are false and rude –. It is self-evident that in Nature we have to find ourselves, and it is clear that science will not be created until we have found ourselves in it. You say that in science it all comes down to experience; and that is fine: on the condition that it will be a double experience, one external and one internal, and both united to a point that they form a single one. But this is, for you positive thinkers, materialists and animists that you are, absolutely impossible not only to do, but even to conceive, because you may well see things one after the other, but you lack the ability to grasp two things at once; and you are good at looking from the outside, but from inside – in media res – you'll never be able to deal with it.]

De Meis's theoretical identification of an external and internal dimension of reality reverberated deeply with those naturalist authors who at the time were experimenting with and elaborating new literary styles and artistic voices. The epistolary form of his novel, furthermore, enhanced the dialogic structure of the book, presented as a conversation between two characters who set in motion a series of philosophical and epistemological exchanges that resonated with current discussions on the novel. Verga, like Capuana, recognized the validity of De Meis's theoretical formulations and embraced his vision of literature as a main instrument of investigation of both the internal world of feelings and the external environment – the only stage where the inner life can surface into consciousness and become material for observation and artistic expression. *Storia di una capinera,* with its narrative at once social and intimate, exemplifies this double mandate of fiction, recognized by De Meis as the true expression of modern subjectivity. The authenticity of Maria's expression – for readers acknowledged her authentic voice – corresponded to Verga's search for connection between the inner world of a character's experience and the external reality in which the character is to act. And even Capuana, who considered Maria a "figura incerta, morbida, senza lineamenti spiccati" [an indistinct figure, without any noteworthy traits], recognized Verga's art precisely in his ability to crystallize the coming together of inner and outer forces, to render Maria's thoughts, voice, and character through a process "diligentemente e acutamente osservato" [diligently and accurately abserved], one in which "l'autore è sparito, è rimasto l'artista" [the author disappears, and what is left is the artist] (Capuana, "Rassegna letteraria" 308)

Giovanni Verga's *Storia di una capinera* reflected the literary debates briefly summarized above as he attempted, through his epistolary novel, to engage the national literary community with themes and ideas that he deemed central not only to his poetics but to the literary discourse of his time. Post-unification Italy, with its main centres of cultural and social life – Florence and Milan – must have generated in Verga, newly arrived from Sicily, a reaction of wonder and fascination (but also of estrangement, which later led to his return, artistically, to his land of origin with his *verista* production of Sicilian narratives). Those first years spent in the north of Italy sparked reflections on the nature of this new world unfolding in front of his eyes, with all its riches of ideas and opportunities, all its various appearances and contradictions; a world he frequented but never completely belonged to.[25] Perhaps it was this sense of anxious attraction he experienced towards modern society – with its hope and promise of purposeful progress – that created the very ambiguity with which artistically he approached his *Storia di una capinera*, infused with elements derived from both his past – his own personal memories of convent life in Sicily – and his present – his awareness that there was a section of public opinion all too willing to condemn certain religious practices and to react emotionally to a story of injustice done to a young girl. The convent represented a remnant of old traditions and social patterns, and to some extent of Verga's own world (not surprisingly he resorted to family memories for his story). But in narrative terms it also constituted not so much a simple contrast to modern life, a term of opposition between what used to be and no longer is, as a symbolic (and thematically still effective) reminder of the constraints (old and new) imposed by society on the individual's personal aspirations. The theme of Verga's epistolary novel, involuntary committal in a convent, was less a reflection on the difficult process of reconciling old and new social conventions than an analysis of the continuing existence of barriers to the fulfilment of human aspirations in a society focused on an idea of progress measured mainly in terms of economic success. As a genre, the epistolary format facilitated the construction of a literary discourse and also fuller communication with the reader, allowing the characters to speak for themselves, creating a more direct relationship between character and reader, and making the author's presence as inconspicuous as possible. The letter form, additionally, gained modern valence for Verga as a symbol for modern communication and provided him, during this initial phase of his literary career, with a narrative strategy that implied the presence of a national public of readers and that therefore revealed his ambition to address the collective Italian, rather than regional, literary identity. The epistolary fiction he produced reflected and contained the world of letter writing he was a part of.

Like many of his contemporaries, Giovanni Verga frequently engaged in letter writing, and during his lifetime he produced a large corpus of letters. Read today, these letters provide valuable information about the author's personal and professional life and also testify to the way in which Verga took advantage of letters for their instrumental as well as creative functionality. But it would be a mistake to consider correspondence a mere reflection of the sender's soul; it inevitably requires a certain degree of artifice, whether conscious or unconscious – the construction of a fictional mask: "l'intenzione insomma di costruire se stessi calandosi in una forma che insieme, proprio attraverso la lettera, si vuole conquistare e introiettare" [the intention in other words to create a persona by adopting a form, the letter, through which one can project and introspect at the same time] (Finotti 92). Reality intermixed with self-invention turns the epistolary narrative into a layered text (a palimpsest, to use Finotti's metaphor) that reveals through the multiplicity of its agency (sender, receiver, and transmitter of the missive) and the ambiguity of its semantic connotation (private and public at the same time) the complexity of the communicative act of letter writing. It is through artifice that correspondence, like fiction, simulates reality and seduces the reader into the illusion of authenticity and the hope of connection. Letter writing – whether in fiction or real life – thrives on narratives of nostalgia where the subject longs to reconnect to an idealized notion of the self, seen through the prism of a dialogical relationship with one's familiar world.

An anecdotal example of how Verga availed himself of the opportunities provided by the letter as a new medium of communication may be found in a missive written on 8 May 1869, in which the Sicilian writer, then living in Florence, encouraging a practice – unauthorized but commonly used – of writing minute encrypted messages inside the strip of paper enfolding a newspaper (which cost less than a letter to mail), explained to his mother the codification of a simple list of expressions to be used for their clandestine communication:

> Ogni giorno, quando almeno lo potrai, mi manderai un giornale di costì o di Palermo. Non dovrai mettere altro nell'interno della fascia che la data del giorno in cui mi spedisci il giornale: sull'indirizzo della fascetta poi metterai quei segni che t'indicai ieri per dirmi quel che vorrai, per esempio il segno sotto Catania, così:

Catania – vuol dire: *stiamo bene tutti;*
lo stesso segno sotto il mio nome, così:
Giovanni Verga – vuol dire: *scriverò colla Posta;*
il segno stesso messo soltanto sotto il *Verga* vuol dire:
Scrivimi e dicci come stai.
Per parte mia farò lo stesso:
Catania intendo dire: *sto bene.*
Mario Verga: *scriverò colla posta domani;*
Sicilia, messo in un angolo della busta *scrivetemi e ditemi come state.*
Non dimenticate questi segni.
Resta inteso che se qualche volta non riceverete il giornale sarà segno che l'avrò dimenticato e non starete in pensiero. (Verga, *Lettere sparse*, 11–12)

[Each day, or as often as you can, you will send me a newspaper from home or from Palermo. You won't have to put anything else on the band except the date of the day in which you send me the newspaper. On the address band you will then place those symbols that I indicated to you yesterday, to tell me what you would want, for example the symbol under Catania, like this:

Catania – means: *we are all well;*
And the same symbol under my name like this:
Giovanni Verga – means: *I will write using the Mail;*
The same symbol placed underneath only *Verga* means: *write to me and tell us how you are doing.*
I will do the same on my part.
Catania will mean: *I am well.*
Mario Verga means *I will write using the Mail tomorrow;*
Sicilia written in the corner of the envelope will mean *Write to me and tell me how you are doing.*
Do not forget these signs.
It is understood that if at any time you should not receive the newspaper, it would be a sign that I have simply forgotten, and you should not worry.]

These codified messages did not completely replace normal correspondence. Verga frequently wrote real letters to his family when he was away from Sicily, and though it is not clear how extensively he and his family resorted to this expedient, the fact that he felt so comfortable in using it reveals the author's familiarity with a system only recently established and his eagerness to take full advantage of all its possibilities.[26] The recently nationalized postal service allowed him, while in Florence or Milan, to maintain a close relationship with his family as well as to develop, while back in Sicily, a national network of professional relationships by writing to some of the main figures of the national intelligentsia with the aim of recreating from a distance those "chiacchierate d'arte" [artistic conversations] that he valued enormously and nostalgically invoked in his letters.[27] Letter writing enabled Verga to imagine having a conversation with his family, or with other literary figures, or, serendipitously, with his readers, in an effort to build a career as an established writer and gain literary recognition. "Capitato oscuro a Milano ritornava famoso in Sicilia" [Unknown when he arrived in Milan he was famous once he returned to Sicily], as Raffaello Barbiera noted in his biographical essay on Verga (347). *Storia di una capinera* played no small role in this initial process of self-invention and self-affirmation. Writing letters, whether for personal or fictional purposes, enabled Verga to create his literary persona and become part of a connective national system of communication that promoted the circulation of ideas and the formation of a collective national and literary identity.

4 Cœur-responding with Her Readers: The Sentimental Politics of Matilde Serao's Epistolary Fiction

Non ne' gabinetti e non negli accampamenti si librano i destini d'un popolo; né tanto possono la frode o la forza, i capricci di pochi o le ciancie. L'anima della nazione sta nelle mani sue stesse: la politica sua s'esercita continua, onnipotente, nella chiesa, nella casa, nel cuore.

[Neither in cabinets nor camps is people's destiny defined; nor through fraud or force, caprice, or the mere words of a few. The soul of a nation is in its own hands: its politics is eternally and almightily expressed in the church, in the home, in the heart.]

Niccolò Tommaseo[1]

Fu quella che voi altri del Mezzogiorno chiamate "la passione," un delirio di lettere, di telegrammi, di fiori, di lagrime, di singhiozzi, di gelosie, una fiammata che salì al cielo.

[It was what you people in the South call "passion," a deluge of letters, telegrams, flowers, tears, sighs, jealousies, a flame that reached the sky.]

Matilde Serao[2]

Matilde Serao's literary career has traditionally been divided into two main periods – the naturalist, and the sentimental or erotic-mundane. The first period, usually more valued by literary critics, dates from 1881, with *Cuore infermo*, to 1890, with *Il paese di cuccagna*, which is often considered her last work in the naturalistic, *verista* vein.[3] In 1891, Serao began the publication of what have been described as "thousands of sterile, inflated, and evanescent pages" (Gisolfi xii) in which the Neapolitan writer described the sentimental life of the aristocracy and the upper middle class. All of Serao's fiction developed a narrative mainly centred on female characters, but while her realistic novels focused on women of the lower classes, her later work – created under the influence of the French writer Paul Bourget[4] – addressed the emotional life of women seemingly unburdened by economic preoccupations. A closer look at her literary production as a whole, however, reveals a much less distinct separation between these two phases of her career, especially when seen through the lens of the author's overall pessimistic portrayal of the realities of women's lives. Her female characters, whether poor or affluent, are never really independent of social or economic constraints, and are shown as being fatally trapped by their limited life circumstances. This is not to say that Serao's fictional production does not present a diversified voice of artistic expression (naturalist or sentimental), but her different voices originate in a unified artistic vision. Nor I am suggesting that Serao's overall pessimism undermined her belief that literature ought to play an instrumental role in the process of social and cultural advancement in Italy. However, while many of her contemporaries, especially the positivists, hailed reform as a main means – spiritual and material – for the cure of Italy's social and economic ills, Serao remained sceptical about the ability of progress to eradicate the problems afflicting the most vulnerable in society, especially women. Whether Serao is describing the meagre lives of people living in the poor neighbourhoods of Naples or the pangs of love suffered by affluent ladies in elegant apartments, a sense of inescapable misery looms over her characters. If suffering were the natural human condition for Serao, art – and, as I will elaborate further, sentiment – constituted the means by which individuals could create a network of spiritual connections within the human consortium.

Serao's poetic elaboration of human suffering is but one variant of what critics have recently come to recognize as a fundamental aspect of Italy's post-1860 culture: anxiety – that is, the expression in post-unification Italy of a national culture infused with strategies and rhetorical discourses denoting the emergence of an anxious modern subject.[5] The emergence of this subject represents the manifestation of a sense of inadequacy resulting from the fast-changing realities of the post-unification period and expressed through different artistic forms: the literary experimentations of the *scapigliati*, the pessimism of the *verista* movement, and, later, the modernist and avant-garde rejections of traditional forms of literary expression. Serao did not necessarily subscribe to any of these schools (although she more than once grouped herself with Verga, Capuana, and De Roberto),[6] but her artistic aspirations stemmed from an epistemological concern with the emergence of new subjectivities and gender identities within the context of Italy's post-unification social and cultural transformation.

Suzanne Stewart-Steinberg, in her analysis of Serao's *La conquista di Roma* (1885), aptly illustrated how Serao, despite the seemingly repetitive plot of this parliamentary novel, managed, "in a confluence of political and aesthetic problems of representation," to render "to culture a new role in the making of subjects" (99). According to Stewart-Steinberg, Serao's novel tapped into a genre that was not only popular in those years but that also expressed a "growing pessimism regarding the Italian parliament's ability to effectively lead the country" (116). This pessimism, however, was not confined to this novel or genre alone, as it pervaded Serao's entire oeuvre. Whether sentimental or *verista* in its inspiration, Serao's fiction exposed the pervasive feeling of confusion and uncertainty experienced both by the individual at the personal level and by Italians collectively during a time marked by social turmoil and a perceived sense of political and social insecurity. Yet, for Serao, it was not only a question of giving voice to a social malaise but also of expressing artistic frustration with the current literary schools. Asserting that "il naturalismo nel romanzo è una forma infelice, inutile o dannosa all'arte" [naturalism in a novel is an unhappy, useless or harmful form for art], Serao admitted in 1883 that "per l'arte e per gli artisti, è un momento pieno di affanno. Mai come in quest'anno trascorso vi è stata maggior lotta interiore, fra i vecchi ideali che ancora resistono e ogni tanto rinascono prepotenti nella coscienza e i nuovi, ancora incerti, ancora fallaci, spesso bugiardi nell'esperimento, ma che si vengono imponendo, come la verità dei giorni moderni" [for art and artists, this is a difficult moment. Never until this past year has there been such an internal struggle between old ideals, which still survive and once in a while resurface to trouble our conscience, and new ones, still uncertain, misleading, often improperly put into practice, but dominating the truth of our modern day existence] (Pupino 208). Serao, a writer who was familiar with current European literary movements and sought new forms of literary expression, interrogated the way in which art manifested itself as a living ideal, "da trasformarsi in carne, ossa colore e vitalità nella propria opera" [giving flesh, bones, colour, and vitality to one's work] (Pupino 208), and strove throughout her literary career to create a fiction that would respond to specific ideals or artistic questions. Seemingly consolatory and reassuring, Serao's sentimental fiction reveals on closer inspection a more complex articulation of current artistic and social concerns than is normally attributed to her later work.

Because sentiment is central to Serao's later literary production, this chapter explores her sentimental fiction – with a focus on her use of the letter, traditionally considered the most natural vehicle for the written communication of inner feelings – and challenges the notion that late-nineteenth-century sentimental narratives, which Serao's epistolary fiction eminently represented, constituted merely a conventional genre aimed at providing mindless entertainment for the expanding reading public. Illusory and ultimately unattainable, love appeared to Serao to be a chimera; in reality, even at the highest level of human passion and dedication to another, the individual remained fatally isolated. Serao's sentimental fiction is filled with stories of doomed love that chronicle the struggles of women trying to reconcile personal aspirations with social expectations. Starting from the premise that modernity and progress were no guarantee of happiness, Serao's sentimental fiction embraced the prescriptive principle, shared by many post-unification writers, that art had a functional and liberating effect – functional insofar as it contributed to the formation of a national consciousness to which women since the Risorgimento had been identified as pivotal contributors in their role as mothers and educators; liberating, because art according to Serao created a space in which the individual could find respite from the misery of life. Her conception of sentimental literature – which could be defined as a "civic pedagogy of the heart" – was in this sense not dissimilar from that of other contemporary authors who, like her, enjoyed great popularity among readers. Edmondo De Amicis famously popularized the image of the "heart" as a central and centralizing narrative organ of the national body politic. And Cesare Cantù, in a book stressing the intimate correlation between heart and common sense (*Buon senso e buon cuore,* 1872), praised the vast educational and moral scope of literature.[7] Serao, like De Amicis and Cantù, strove to turn the emotional force expressed by the fictional text into positive moral sentiments, which were supposed to inspire and, ultimately, build the character of her readers and, by extension, of the Italian nation. Feelings, at least from the point of view of artistic creation, provided Serao with an emotional currency that enabled her to develop a connection with her readers, a shared patrimony of values and ideals. The sentimental fiction of Matilde Serao ultimately offered the promise of personal and collective transformation (through the promotion of a specific set of moral and social values) rather than mere escapism from the realities of everyday life. Her fiction of the "heart" offered a narrative representation of a relational interaction, of an experience filled with love and desire, the story of an affiliation or a membership, which, as Gabriella Turnaturi noted in *Betrayals*, is necessarily formulated within the framework of a "we" that "brings with it the possibility of betrayal, separation, or rupture" (3). The sentimental fiction of post-unification Italy represented the narration of an Italian melodrama, a love story with an idealized notion of cultural unity, filled with hope, ruptures, separations, betrayals, tears, and anxieties. Matilde Serao became a main interpreter of the hopes and anxieties that accompanied Italy's effort to create a national cultural identity during the post-unification period.

In an exchange of articles published in 1880 in *Il Corriere del Mattino*, Serao explained to the director of the newspaper, Martino Cafiero, what she meant by "la politica del cuore" [the politics of the heart], an expression she used to define both the emotional import of art and women's sentimental approach to politics. Juxtaposing the political ideal of the monarchy to that of the republic, Serao saw the former as the symbol of both woman's condition in society ("le hanno troppo detto che è regina in casa sua, ha troppe volte detto ad un uomo che egli è il suo re, perché ella possa amare qualche cosa di diverso che la monarchia" [they told her too often that she is the queen of her home, she has too often told a man that he is her king, for her to love anything else but the monarchy]) and, more generally, of the human experience. As Serao explained to Cafiero:

> Allora come ella dice nella sua profezia umanitaria, nessun re, nessun soldato, nessun reietto – ovvero tutti re, tutti soldati, tutti felici. Poniamo? È lutto per quel Tempo, Signor Cafiero, perché "l'Arte sarà morta." Tutto sarà eguale, tutto sarà livellato. Tutto si rassomiglierà. Cesseranno allora tutti i drammatici contrasti che sono la tela della vita. Cesseranno la loro artistica e feconda opposizione, l'ombra e la luce, lo spirito e il corpo, la forma e l'idea, Mefistofele e Margherita, la lagrima ed il sorriso [...] la reggia e il tugurio, la passione della terra e la passione di Dio. In questo magnifico mondo, Signor Cafiero, non vi sarà modo di amare, di soffrire, di vivere.[8]

> [So, as you state in your humanitarian prophecy, no king, no soldier, no outcast – or rather, all kings, all soldiers, all joyful. And if that were to happen? That time will be mournful, Signor Cafiero, because "Art will be dead."

> Everything will be identical, all will be levelled off. Everything will appear the same. All the dramatic contrasts which make up the very fabric of life will cease to exist. Their artistic and fruitful opposition, the shade and the light, the spirit and the body, the form and the idea, Mefistofele and Margherita, the tear and the smile [...], the palace and the shack, the passion for the land, and the passion for God, will all cease to be. In this magnificent world, Signor Cafiero, there won't be a way to love, to suffer, or to live.]

Strongly infused with pessimism, Serao's politics of the heart stemmed from her belief that suffering constituted an essential and inescapable human condition – and a creative force. Social injustice and human frailty could not be avoided in life; but they could at least be alleviated through the cathartic power of art. In an interview with Ugo Ojetti in 1894, Serao further elaborated on the relationship between art and suffering:

> Ora guardiamo al popolo: null'altro risponde in lui a quei nostri sentimenti che una irrequietudine continua, pungente, una aspirazione all'ideale fuori dalla faticata vita di tutti i giorni. E le difficoltà economiche sociali sono la causa determinante di quella inquietezza. Esso non sa dove dirigersi per il conforto, noi dobbiamo mostragli la via; ed è opera santa, ed è opera cara, ché, confortando noi, confortiamo esso pure. Il soffrire ci unisca, noi e il popolo sentiamo comuni il soffrire, sentiamo sui colli nostri lo stesso giogo dell'umanità greve, della tristezza animale, della caducità. In questa comunione di dolore l'amore ci unisca [...] con questo non scioglieremo la questione sociale, ché ciò è impossibile, ma la leniremo. (Ojetti 241)
>
> [Now, let us look at the people: nothing else resonates emotionally in them but a continuing stifling restlessness, an aspiration to an ideal removed from laborious day-to-day life. The [people's] economic and social difficulties are the cause of this restlessness. They know not where to turn for comfort, we must show them the way; and this is a sacred and dear task, that, in comforting ourselves, we comfort them as well. The suffering unifies us, we and the people feel a common suffering, we feel the same burden of a grave humanity upon our necks, of primordial sadness, of frailty. In this communion of pain, love unites us [...], we will not resolve the social question, which is impossible, but we will lessen it.]

Serao took a pessimistic view of the possibility of overcoming social inequalities and maintained a conservative stance towards legislation promoting gender and/or class equality, instead envisioning artistic creation as an instrument with which to improve society, but at the level of moral conduct rather than political action.[9] And while, in her answer to Ojetti in 1894, Serao was referring to the spiritual aspect of literature and to the need not to reduce realism to mere materialism, her interpretation of human suffering within a context of artistic production also extended to the sentimental fiction that she began to write at that time and must be viewed as the overarching creative idea that inspired all her fiction. Already in 1877, complaining about the pervasive positivistic approach to art, Serao contrasted the harshness of reality to the liberating effect provided by art through "la poesia, i fiori, e l'amore" [poetry, flowers, and love].[10] A few years later, in 1906, she further clarified this point when, in a pamphlet titled *Sognando* [Dreaming], she insisted on the transformative power of art, which, with its "misterioso elisir che è la fantasia" [mysterious elixir which is fantasy], is able to make life seem more magical than it is. Fantasy or "sogno" [dream], as she conceived the process of writing, becomes for Serao a productive act ("il sogno diventa operoso" [the dream becomes productive]) because of its extraordinary capacity to transform society ("forza capace di sollevare montagne" [a force able to raise mountains]).[11]

Serao's sentimental fiction mirrors, and thus contains, the harshness of the reality more explicitly portrayed in her early writings, but focuses on the theme of love for both its emblematic function (she thought marriage and relationships were the two central factors in women's lives) and its cathartic artistic value. Many of Serao's love stories portray women who secretly write to their lovers and read letters from them. A sense of ambiguity, embodied in the literary representation of an epistolary exchange, is maintained by Serao, as the letter here functions as the vehicle for both the expression and the censorship of female emotions (women express their hidden desires only in the intimate discourse of letter writing). This double representation of female sexual desire has inspired different critical interpretations of Serao's sentimental fiction. Most critics have underlined the writer's essentially conservative view of women's sexual behaviour, while others, such as Laura Salsini, have seen in Serao's love stories an "alternative subversive reading" (75) of female erotic desire. My study of the epistolary form of these stories makes the conservative interpretation more compelling, because the structure of the texts forces us to read her narrative as a construction of an essentially didactic discourse, framed within the general national project of nation making. The reader cannot avoid identifying with the fictional receiver of the letter, and the letter thus becomes a form of "authorial constraint," as explained by Margherita di Fazio, who, in her analysis of epistolary communication, demonstrates that the authorial "I," "seguendo i propri desideri o le proprie necessità, scegliendo il momento per sé più adatto, scrive a un tu, il quale riceve, in un tempo che non ha scelto, una lettera che ha il dovere di leggere e che lo obbliga alla risposta. La lettera perciò, quando apre lo scambio epistolare, rappresenta una costrizione" [following one's desires or needs, choosing the best moment, writes to a you, who receives in a time not chosen by him a letter which he has the duty to read and which he is obliged to reply to. The letter, therefore, when it opens an epistolary exchange, represents a constraint] (9).

Like all prescriptive literature, Serao's epistolary fiction is an act of "authorial constraint," mitigated, however, by a sense of familiarity conveyed by the epistolary activity – to "cœur-respond" with her readers, in other words to communicate through the affective force of the sentiments expressed in the letter.[12] Serao's use of the epistolary form for her sentimental fiction reflects her urge to influence her audience and, to this end, her willingness to tap into the life experience of her readers who, with the nationalization and larger diffusion of the postal services, were now availing themselves of this form of communication more frequently than ever before. As Umberto Eco has pointed out, "attraverso un romanzo scritto come serie di lettere il lettore dell'epoca trovava qualcosa di familiare, capiva che si stava parlando della propria sfera intima [...] E allora capiremo che non è un caso se molti dei primi romanzi sono romanzi epistolari" [through a novel written as a series of letters, the reader of that time found something familiar, understood that they were speaking of her own private sphere] (7).[13] Serao's epistolary fiction belonged to the later period of her career, but the author was at this point nevertheless exploring the formal properties of the literary text, at a time when the novel as a literary genre was still struggling to find its place not only in the literary canon but also, more generally, within Italian culture, since the majority of the novels published in Italy were not written by Italian authors. Writers, especially those who aimed to engage as large an audience as possible, were forced to face questions related not only to the production of the literary object (i.e., What literary strategy, language, or style is appropriate for the modern novel?) but also to the reception of it (i.e., What strategies would be most effective in engaging the reader? How can a text be made relevant to a readership only recently introduced to literacy and literature?). The epistolary format allowed Serao to address the issues of both formal textual properties and readability while affording her an effective means of communication with her readers, who were by then increasingly using letter writing not only for personal but also for public exchanges of information.

In one of her short stories, included in *La vita è così lunga* (1918), Serao described, in minute and leisurely detail a scene of epistolary life with a woman as its protagonist. The story "Livia Speri" opens with Livia, the protagonist, reading a book by Paul Bourget, when she is handed a letter by her servant:

> Livia Speri lasciò il libro che leggeva, *Les detours du coeur* di Paul Bourget, fra le mollezze dei cuscini della sua *chaise longue*, prese, dal vassoio di bronzo giapponese, la lettera che il servo le porgeva e la guardò, un istante, coi suoi grandi occhi azzurri-violaceo [...] ancora un momento la lettera rimase nelle sue belle mani, [...] con un sottile coltellino dalla lama d'oro, ella tagliò direttamente la busta. Estrasse un fascetto di fogli lievi [...] e tenendo la lettera in una mano [...] ella lesse, pianamente, lentamente quanto le scriveva Roberto, il suo amante. (*La vita è così lunga*, 45–6)

> [Livia Speri left the book she was reading, Paul Bourget's *Les detours du coeur,* among the pillows on her *chaise longue,* took the letter that the servant had given her from the tray made of Japanese bronze, and looked at it for an instant, with her big blue-violet eyes [...] for a moment longer the letter remained in her beautiful hands, [...] with a thin letter opener with a golden blade, she cut open the envelope. She extracted a small bundle of flimsy sheets of paper [...] and holding the letter in one hand [...] she read, calmly, slowly, all that Roberto, her lover, had written to her.]

Here, the function of the letter is twofold: it represents a social practice (a woman receiving and reading a letter), and it helps the reader identify with the fictional character. Ultimately, the letter is used to enhance the exemplary and didactic tone of the fiction, and, not surprisingly, writers who use the letter as a literary strategy tend to write a fiction that is highly affective, meaning that it is meant to influence the reader towards specific ideological positions and moral values. Foscolo's *Ultime lettere di Jacopo Ortis,* to cite a famous example of the epistolary novel in Italian literature, succeeded in making the reader identify with the protagonist to such a degree that many young Italians attempted suicide in emulation of their idealized Jacopo. Decades later, Foscolo's novel continued to captivate readers, especially Italian youth, as attested in Serao's novel *Romanzo della fanciulla,* in which a passionate young woman, Alessandrina Fraccareta, is shown surreptitiously reading *Ortis* while nearby her friend Teresa is writing a letter. Serao, who often spoke about her correspondence with her readers, understood the strategic effectiveness of the letter as a form of communication, and adopted it both to enhance the authenticity of her fiction (shaped by the realities that emerged from her correspondence with readers) and to promote discussion of specific issues or ideas among her female readers. Influenced by the author's familiarity with letter writing as a social practice and also by her belief in the cultural potential of epistolary communication, Serao's sentimental fiction tapped into the properties of postal culture, of a growing community who used the letter as an important means of cultural exchange and information distribution.

Between 1890 and 1918, Serao published many short stories written in epistolary form. They appeared in collections such as *Fior di passione* (1890), *Gli amanti* (1894), *Lettere d'amore* (1901), *Novelle sentimentali* (1902), and *La vita è così lunga* (1918), all of which focused, as their titles suggest, on passion and love intrigues. Letters were also included in the novels *Tre donne* (1905) and *Ella non rispose* (1914) (the latter written entirely in letter form), both of which resembled Serao's other novels in thematic content and narrative strategy. The plots are quite repetitive and narrate, with very small variations, the downfall of women who seek fulfilment of their sentimental and erotic desires outside marriage. Having married, more often than not, for economic reasons, these female characters find themselves emotionally isolated, resorting to adulterous love in an attempt to make up for the emotional emptiness of their marriages. Structurally and stylistically the conversational tone and low register of the language, the relatively self-enclosed, sequential configuration of the letters, and the detailed descriptions of the setting reflect Serao's journalistic writing, with its descriptive, synthetic, linear narration of events. All of Serao's sentimental stories are presented as exempla, as warnings to readers not to project their fantasies beyond the fictional space of the literary text. In this sense the fantastic world is a space of expression of an inner desire that cannot be satisfied in real life, and while Serao does not delve too deeply into the psychology of her female characters, the letters offer a fragmentary glimpse of a holistic vision of female identity – a representation, that is, that takes into account both the inner and the outer realities of women's lives.[14] The repetitiveness of the sentimental plot is metonymically represented by the succession of letters in her fiction.

In "Lettera d'amore," published in her 1902 collection, *Novelle sentimentali*, Serao wrote,

> Mia diletta Nice,
> quanto mi è difficile scriverti questa lettera! E dire che ve ne ho scritte tante! Malinconiche, stupide, iraconde, amare liete perfide: lettere di ogni genere, poiché riflettevano fedelmente lo stato bizzarro del mio spirito: lettere, cioè ineguali, strane capricciose che vi possono aver sorpreso, commosso, indignato, o Nice, ma che, sempre, confessatelo, vi debbono aver dato impressione sincera dell'amore.[15]

> [My beloved Nice,
> how difficult it is for me to write you this letter! And to say that I have written so many to you. Melancholic, stupid, irritable, loving, happy, deceitful: letters of every type, so long as they faithfully reflected the bizarre state of my spirit. Letters, that is to say, unequal, strange, capricious, that may have surprised, touched, made you angry, O Nice, but that always, confess it, must have given you the sincere impression of love.]

The opening of this letter represents a sort of manifesto of Serao's sentimental fiction. She admits that her many letters are heterogeneous in their moods, but points out that they are instrumental in allowing her to describe the human emotions that inspired her to write them. It is on the emotional value of the love story that Serao constructs her fiction, and it provides her defence against the detractors of a genre – sentimental fiction – that was considered trivial and evanescent. Serao defended her love stories in the introduction of her epistolary novel *Ella non rispose*:

> Amico lettore, ti prego sii sincero. Quando i tuoi occhi curiosi si saran fermati, un istante, sulla copertina di questo volume e l'avran letta, non ti abbandonare a quel gesto tutto convenzionale, tutto esteriore, per cui un romanzo di amore fa levare le spalle, in atto d'ironica noia e di sazietà beffarda. Non levar le spalle, caro lettore: non imitar quello che tanti altri esseri umani faranno, come te, imitandosi fra loro: non obbedire a una formola superficiale e che ti è estranea: abbi il coraggio di essere sincero. Cedi a quella intima, segreta nostalgia sentimentale, per cui ogni istoria di amore, la più umile, la più comune, attira me, te, gli altri, come un fluido sottile: cedi a quel bisogno di conoscere, ancora, un altro caso di amore, bisogno che tutti sentiamo nell'anima, bisogno mai, mai, interamente soddisfatto. (v–vi)

[Dear reader, I beg you to be sincere. When your curious eyes have rested, for an instant, on the cover of this volume, and have read it, do not abandon yourself to that conventional gesture, only for show, shrugging your shoulders at a love story in an act of ironic boredom and mocking satisfaction. Do not shrug your shoulders, dear reader: do not imitate what so many other human beings will do, like you, mimicking each other: do not obey that superficial example that is alien to you: have the courage to be sincere. Give in to that intimate and secret sentimental nostalgia through which every love story, the most humble, the most common, attracts me, you, and everybody else like an imperceptible fluid: give in to that need to know yet again another love story, a need that we all feel in our souls, a need never, ever, fully satisfied.]

Love stories have for Serao a universal appeal; that is, they appeal to a universal capacity to feel love but also to an eternal desire to explore and understand the meaning of love. Once a reader is able to overcome the common prejudice against sentimental novels, Serao suggests, she will find that they are not simply vehicles for self-indulgence in emotion but, on the contrary, offer a vision of human desire.

Here it is useful to elaborate on what makes emotion, passion, and sentiment different from one another, not only in the context of Serao's fiction but, more generally, in post-unification Italian sentimental novels. Before the Enlightenment and the rise of the positivist and scientific methods, emotions and passions were considered to be the same – natural, affective reactions to sensations of pleasure and pain; however, during and after the eighteenth century a distinction arose between emotions and passions. Within medical discourse, the latter, especially with the development of the disciplines of psychiatry and psychoanalysis, acquired a negative connotation, to the point of being considered part of a mental pathology, a state of folly; while emotions began to be studied and viewed by scholars and scientists for their cognitive and moral value (Pulcini, "Breve storia delle passioni" 178–9). More recently, emotions have become an important focus of investigation in history, psychology, sociology, philosophy, and cultural studies. In these studies, emotions "have come to be seen as not simply individual and inner phenomena, but as collective cultural and historical experience, potentially eroding dichotomies such as inside/outside, individual/collective, and private/public" (Harding and Pribram 2). In a recent issue of *Critical Quarterly*, for instance, Patrizia Lombardo pointed out how scholars, especially in analytical philosophy, have in the last twenty years demonstrated that emotions "have motives or reasons" and "influence the rational pattern of decision-making," and that "affective rationality and irrationality interact" (1, 2). Thus, according to Lombardo, emotions are essential to decision making. But also, citing Aristotle's *Metaphysics*, she suggests that "the discovery of affective rationality throws light on the relation between emotions and many kinds of knowledge [...] in particular, literature is a form of knowledge, knowledge of the affective life" (7). And Alberto Mario Banti, in an essay included in a volume devoted to emotions and politics in Italian history, points out a fact that it is useful to keep in mind when thinking of the kind of readers – not highly educated women – who made up Serao's public: that is, how emotions are particularly efficient as a communication tool, especially in cultures where "molti non dispongono che di assetti cognitivi fragili, quando non sono del tutto analfabeti e scarsamente allenati ad affrontare percorsi concettuali complessi" [many have only fragile cognitive tools, if not being all together illiterate or scarcely used to understand complex conceptual processes] (43).

Such a taxonomy of feelings and the redefinition, in more positive terms, of the affective role of emotions, warrants further elaboration, particularly with reference to the way in which late-nineteenth-century popular or sentimental fiction used sentiment, distinguished here from emotion, normatively to regulate inner feelings that, after the unification of Italy, were seen as instrumental in the creation of a national Italian culture based on a presumed core or "heart" of a shared national identity. While passions, as noted above, constituted an innately destructive inner force, and emotions were the natural expression of affective reactions that could have a more or less positive outcome (Serao's creative notion of artistic inspiration, for instance), sentiments were envisioned as culturally constructed and organized around social and cultural relationships. They provided, in other words, a set of feelings that could be conditioned or managed in order to foster the creation of an acceptable new social order. Serao's sentimentality, especially when seen in the context of a post-unification civic pedagogy of nation formation, offered an "organized" set of narrative strategies, or sentiments, aimed at creating a pedagogy of cultural and social unity. Serao's portrayal of sentiments suggests that feelings operate relationally and, when given a positive direction, can foster consensus in society. Because representing feelings necessarily involves an ethical judgment (the performative aspect of feelings always assumes the presence of a real or imagined spectator to the joy, suffering, or pain expressed), sentimental fiction requires its readers to become interpreters of the values portrayed in the narrative. Critics who theorize about sentimentalism suggest that "the sentimental text is a rhetorical construct whose aim it is to affect the reader, to move the reader – *movere* in the classical terminology – by means of pathos" (Herget 4), and that "sentimentalism declares its project to be the valorization of feeling, and of the communicability of feeling, as the basis of a regeneration of human society" (Denby 83). As several critics have pointed out, the word "sentimental" initially had a positive connotation. For Samuel Richardson, one of the first to use the word, "sentiment" was synonymous with moral feeling (Herget 2); while for both David Hume and Adam Smith emotions were an essential component of the human condition, integral to the social fabric (Evans). Towards the end of the eighteenth century, however, the word ceased to have the positive meaning assigned to it during the Enlightenment and became "a term of near abuse referring to mawkish self-indulgent and actively pernicious modes of feeling" (Bell 2). This is still true today; "sentimental" is still used negatively to describe "a sense of false consciousness based on an indulgence of insincere or delusional emotional attachment" (Nelson 2). The Italian word "sentimentale" also underwent a shift in semantic value during the nineteenth century. In the 1869 *Dizionario della lingua italiana,* compiled by Niccolò Tommaseo and Bernando Bellini, the definition of "sentimentale" was "parola che viene da fuori" [a word coming from outside] and "persona che dimostra di delicatamente sentire gli affetti" [a person who shows delicacy of feeling]; however, a few decades later, in 1908, Policarpo Petrocchi's popular *Novo dizionario universale della lingua italiana* defined "sentimentale" as "che accenna a un sentimento romantico, floscio, affettato" [denoting a romantic, soft, and affected feeling]. Serao was undoubtedly aware of the negative associations of sentimentalism during the late nineteenth century; nevertheless, human sentiments are portrayed positively in her fiction, as the source of that emotional currency she deemed necessary for connecting with her readers. The sentimental letter serves as a means of communicating with her readers, enhancing the author's effort to create a bridge, a narrative space of commonality with a readership envisioned as the recipient of the message of the novel. When Serao in her introduction to *Ella non rispose,* wrote

> Bastava, a me, questa comunione spirituale, questa comunione sentimentale, dovuta solo alla mia emozione umile e sincera, nel segnare le mie storie d'amore e di dolore, dovuta al misterioso e al possente vincolo delle anime, innanzi alla vita, alla verità e alla poesia. E sempre, questa lettera mi è bastata, per dire che la opera mia non era stata un vano e sterile esercizio di letteratura: ma qualche cosa di semplice e schietto, nella sua forza di sentimento. (xii)
>
> [It was sufficient to me to have this spiritual, sentimental communion, springing from my humble and sincere emotion, when recording my stories with love and pain, due to the mysterious and powerful connection existing between souls in life, in truth, and in poetry. And always, this letter is sufficient to say that my work was not a vain and sterile literary exercise, but something simple and frank, in its sentimental force.]

she was defining the sentimental letter, and art in general, as that space of the imaginary in which people could meet and connect through shared emotions. In other words, art has the transformative power (as described in *Sognando)* to connect the inner world of desire or emotion to the outer world of social interaction or sentiment. The artistic space of fictional production is, basically, where the two dimensions of the inner life – emotion and sentiment – meet and are transformed into a positive force.

If emotions – happiness, fear, rage, and so forth – were central to Serao's portrayal of human desire, her sentimental narratives sought to contain them and bring them under the control of rational and social norms. Emotions were thus interpreted as the raw material, the fabric, of human desires, while sentiments constituted the outcome of specific choices made in life that reflected a shared patrimony of values. The ruinous effects of uncontrolled emotions and passions is thus contrasted to the salvific impact of sentiments, valued for their social acceptability and moderating influence on potentially disruptive inner emotions, the strenuous but necessary "esercizio d'automaschermento" [exercise of self-concealment] identified by Vittorio Roda as a central element of Serao's representation of her female characters (318). "Popolana, borghese o aristocratica che sia, la fanciulla adotta una precisa strategia, quella dell'occulatarsi, del mascherarsi, del rendere impenetrabili, sono parole della scrittrice, i misteri del suo spirito" [Whether from the lower, middle, or upper class, the young lady adopts a precise strategy, one of self-concealment, of masking, of making impenetrable – these are the author's words – the mysteries of her spirit] (319). Serao's sentimental narratives deal with the conflict between the inner impulses of the heart (emotions or passions) and social expectations (sentiments) and provide a filter, an instrument that allows tumultuous emotions to be reconciled with the reality of one's social relations and obligations. Good sentiments and virtuous behaviour appear as the exclusive and unavoidable path to social respectability – for Serao, the only way in which women could attain, if not happiness, at least social acceptance and stability. The stories narrated in her sentimental fiction are a repetition of this single axiom of female wisdom: "Aspra è la battaglia nella vita feminile, ma il motto sconfortato di Giobbe è fatto per la fanciulla" [Women's life is but a hard battle, and Job's disheartened words are made for women] (*Il romanzo della fanciulla* 4). Short story after short story exposes with monotonous predictability the unhappy destiny of women, whose suffering exemplifies the general human condition. Influenced by the author's Catholic vision of human suffering – redeemable only in the afterlife – Serao's conception of love is one that is virtually unattainable in real life because it aspires to a sense of wholeness that does not belong to the human experience. That wholeness, however, may be recaptured, if only momentarily, through art and the cathartic power of the artist, who can make others see things in a new perspective. The epistolary format of Serao's sentimental fiction, furthermore, illuminates both the fragmentary aspect of the narrative experience of love and the need for a message (of love or otherwise) to be delivered. *Ella non rispose,* for instance, presents a compilation of more than ninety letters from an impassioned Paolo to Diana. The love story is seemingly presented as a monologue addressed to the female protagonist, who, forced by circumstance to marry an elderly, wealthy Englishman, never responds to Paolo's letters and silently ponders the misery of her state. But suddenly the correspondence ceases – Paolo stops writing to her and withdraws to an isolated and undisclosed place – and her husband unexpectedly dies in a mountain accident. She then begins to reread Paolo's letters in a final desperate attempt to recapture a last, illusory moment of fulfilment in love. But all is in vain; even her last plea to her maid to find him and tell him of her love is of no avail. That idealistic coming together of two loving souls will never happen. The long sequence of letters and the emphasis on the process of reading and rereading the same letters highlights the role of the reader as an interpreter of the epistolary message, which in this novel is invested with a redemptive rather than a merely consolatory value. Diana dies; the heartbroken Paolo has disappeared; there is no consolation for the reader. What the story offers is a bitter but valuable lesson on how to read correctly the burgeoning popular literature aimed at female readers and usually consisting of simple narratives of sappy romantic love with an "and-they-lived-happily-ever-after" ending. "Perché prendete la vita come un romanzo da camera, dove *amor* fa rima con *ognor*? E con *dolor*?" [Why read life like a domestic novel where love rhymes with every moment, and pain?] Serao wrote in *Tre donne*, another story of hopeless love, involving Don Francesco, a handsome but emotionally cold nobleman, and three women, two of whom, Daisy and Donna Clara, exemplify in their physical attributes different national approaches to gender definition (Daisy, an English woman, is fair and blonde, while Donna Clara, an Italian, is a brunette). Here, Serao taps into a still-pervasive romantic idea that nationality defines character, particularly that of women.[16] Ugo Foscolo, for instance, explored this theme in his pamphlet *Le donne italiane*, published first in England in 1826 and later translated into Italian, describing the sorry state of Italian women of the upper classes, who, unlike their English counterparts, were frequently forced by social convention into uncongenial marriages and encouraged to practise the morally questionable *cicibeismo*. Seemingly accepting these stereotypes, Serao portrays Daisy as a free spirit with an independent income that allows her to avoid a marriage of convenience. Clara, instead, like all young Italian women of her social class, is forced (like her mother before her) to marry an affluent older man. And, yet, in spite of their different upbringing, temperament, and economic condition, these women both suffer heartbreak, deluding themselves about the possibility of marrying Don Francesco and thinking of marriage strictly in terms of romantic aspirations. "Non avevano altro segreto desiderio, o palese," Serao wrote, "che il matrimonio" [We had no other secret or apparent desire but that of marriage] (29). But marriage with Don Francesco is unattainable, since he will not commit to either one of them. What these women are left with, in a dramatic narrative turn, is either death (Daisy commits suicide), or dissimulation, the masking of inner emotions that turns disillusioned women into their own silent confidantes. As Donna Clara writes in a final letter of separation from Don Francesco:

> Ma il silenzio e il riserbo e la compostezza e la freddezza a cui siamo obbligate noi ragazze – né forse potrebb'essere diversamente – ci isolano:

e avendo l'apparenza di gentili bambole indifferenti, insensibili, noi viviamo profondamente una vita interiore tumultuosa, spesso che ci esalta. Chi supporrerebbe che Donna Clara di Nerola [...] che sposa volentieri il vecchio principe di Schillingfurst, chi supporrerebbe che in questa bella ragazza ma esteriore, come tutti dicono, ci sia una passione divoratrice, una passione prepotente e indomabile, che ne ha già distrutto e domani potrebbe distruggerne la virtù? (*Tre donne*, 98–9)

[But the silence, reserve, composure, coldness into which we are forced from childhood – it could perhaps not be otherwise – isolate us. With this appearance of indifferent and insensitive dolls, we live profoundly an interior life that is tumultuous and exalting. Who would guess that Donna Clara of Nerola ... who happily marries the old prince of Schillingfurst, who would ever guess that within this young woman, as they say, beautiful in her appearance, there is a devouring passion, a strong passion that is untamable, that has already destroyed and could again destroy her virtue?]

Marriage, female independence, and economic stability are some of the main themes presented in this novel as well as in the rest of Serao's sentimental fiction. Serao's journalism dealt with similar subjects (sometimes in the form of a letter or a response to a letter from one of her readers or a fellow writer) as she focused both her non-fiction and her fiction on ideas and topics that she deemed of interest to her readers. Women's ideas about marriage were of particular interest to Serao, not only for ideological reasons related to gender-role definitions within the post-unification bourgeois rhetoric of the family but also as a vehicle for her creative exploration of the subtle tensions between the inner and outer realities of women's lives. "Falso in scrittura" is a case in point. The short story, consisting of three short letters, describes the female protagonist's difficulty in reconciling her sense of social obligation to remain loyal to her unhappy marriage with her inner temptation to experience passionate love and engage in an illicit affair with a friend of her husband. The letters describe the different stages of a conventional narrative of female redemption (the wife betrays her husband but in the end repents and returns to her domestic duties), while also exposing the fallacy of conventional love stories by denying a happy ending. The intimate, amorous discourse is contradicted at the very end by the narrator, who warns the reader not to believe everything she has read.[17] The communicative aspect of letter writing, the didactic message with its moral underpinnings, prevails over the intimate tone of the amorous discourse. Serao does not simply describe the protagonist's fall into sin, she warns the reader about its consequences.

Fear of social disgrace seemed to be a source of great preoccupation for Serao, who made a career out of providing her readers with advice on how to embrace proper social conduct, as if behaving inappropriately were a mistake that people could not afford to make. In the introduction to her conduct book, *Saper vivere*, she wrote:

> Ed è molto bene per te, amico lettore, che tu, per tuo istinto di equilibrio, per natural gusto eletto, conosca questo saper vivere, e che, in qualunque ora della tua vita tu non commetta mai uno di quegli errori di condotta, di misura, di scelta che sembrano piccoli ma che, talvolta, portano delle conseguenze meno lievi e, forse, gravi. (vi–vii)
>
> [It is good for you, my dear reader, that by instinct of balance, by natural taste, you know how to live and you know how in any moment of your life not to make one of those mistakes in conduct, in judgment, in choice which might seem little, but sometimes bring consequences that are less little and, perhaps, grave.]

"Appropriateness" as a constructed system of social norms by which to avoid any "faux pas" constitutes a major theme in Serao's writing. In her sentimental fiction, in particular, she uses the juxtaposition between bad emotions and good sentiments to instruct her readership about what she considered, within the context of bourgeois values, appropriate personal and collective behaviour. *Lettere d'amore*, for instance, opens with a narrative excursion through the fashionable centre of Rome:

> Mio amore,
>
> È solamente un'ora che ho lasciato il delizioso nido d'amore, preparato dalla vostra passione e dal vostro gusto eletto per accogliere la divina donna, come voi umilmente e gloriosamente mi chiamate, e come tanto io mi compiaceva, di essere chiamata. A quell'angolo bizzarro, fra via Gregoriana e la scala della Trinità dei Monti, quell'adorabile scala, donde tanti miei sogni si sono evaporati nel cielo crepuscolare di Roma, la piccola casa, alta, guarda, dai suoi balconi, uno de' più belli e più fantasiosi spettacoli del mondo, dai cipressi fatidici di Monte Mario alla cupola di S. Pietro, ergentesi pallida sul pallore delle nebbie che salgono dal fiume, dalle rosee aurore bagnanti di luce l'obelisco della Vergine, in piazza di

Spagna, ai tramonti cinerei dietro gli alberi di quel giardino chiuso che è Villa Medici.[18]

[My love,

It has only been an hour since I left the delightful love nest, prepared by your passion and by your superior taste in welcoming the divine woman, as you humbly and gloriously call me, and as I too was so pleased to be called. At that odd corner, between via Gregoriana and the steps of the Trinità dei Monti, those adorable steps, where so many of my dreams have evaporated into the crepuscular Roman sky, the small tall house, from whose balconies one can see the most beautiful and fantastic views in the world, from the fateful cypress trees of Monte Mario to the dome of St Peter's, palely emerging from the fog that rises from the river, from the rosy dawn which bathes the obelisk of the Virgin in Piazza di Spagna, to the ashy sunsets behind the trees of the closed garden that is the Villa Medici.]

By providing a paratactic description of the familiar elegant streets of the capital city, Serao accompanies the reader on a narrative excursion only seemingly aimless and frivolous. Theoretical elaborations on the trope of flânerie and nineteenth-century visual practices are helpful here in revealing the way in which Serao constructs her sentimental narrative in terms of visual pleasure while emphasizing the immobility of the subject. *Lettere d'amore* epitomizes Serao's general approach to her sentimental epistolary fiction, which reflects what critics have defined as a "panoptic system," that is, the mechanism through which specific subjectivities are created by way of established visual control over society (Friedberg 17). In *Lettere d'amore,* the reader is taken on a visual tour of famous sights of the eternal city but is not allowed to roam independently and never leaves the enclosed space of the domestic realm from which the reading is supposed to take place. As Walter Benjamin famously suggested, "literature is also socially panoramic," and if reading allowed women to step out of the confines of the household, if only in their imagination, the cultural phantasmagoria of the nineteenth century, which according to Benjamin was a peculiar trait of this era, brought also disenchantment and ultimately alienation, referred to by the critic as the gaze of the "alienated man" (97, 104).

The protagonists of Serao's sentimental letters experience a similar state of domestic and public alienation (for Benjamin the public space of the arcade and/or department store and the home belong to the same phantasmagorias of the interior), although they cannot truly be considered *flâneuses*. As Anne Friedberg explained in a study of late-nineteenth-century visual practices, the trope of flânerie can only really be used in reference to the male observing subject; one can speak of a *flâneuse* only when women are free to roam the city on their own without an escort, and this certainly did not happen in Italy until much later. Sufficient proof of the perils of female wandering is provided by the trouble that Checchina, in Serao's eponymous novella, went through in order to leave the house by herself and reach, in the end unsuccessfully, the destination of her romantic escapade. But if in the nineteenth century women were not yet able to move around freely without putting their respectability at risk, the question of female mobility in the public space was crucial in debates of the time about how to reconcile the conflict between tradition and modernity in the process of national social transformation. On the one side, bourgeois society promoted a sense of public decorum, based on a traditionally gender-divided society configured in terms of a male public space versus a female domestic realm, while, on the other, it pursued economic advancement through a cultural and commercial consumerism that not only targeted women but encouraged them to escape the temporal and spatial constraints imposed by tradition. The late-nineteenth-century establishment and growth of department stores, packaged tourism, museums, and illustrated travel literature created, as Friedman explained, "fantasy worlds for itinerant lookers" (37) while at the same time relying on "spectator immobility" and thus offering "a visual excursion and a visual release from [the] confinements of everyday space and time" (28). In addition, as Cristina della Coletta illustrates in her *World's Fair Italian Style*, cultural practices, whether related to popular or higher forms of artistic production, provided the burgeoning cosmopolitan audiences of post-unification Italy with an "exposition" – "the symbolic space where societies offer a spectacular *mise en scène* of their beliefs, fashions, politics, economic systems, and power structures" (6). Serao's epistolary narratives may be considered part of this post-unification cultural spectacle, which, from the pages of the proliferating print media, enticed the newly literate citizens of Italy to become part of the project of nation making.

But this exposure to the gaze of the world represented in the end the very embodiment of the constraints faced by the modern individual, who remains stranded, and, ultimately, alienated on the "threshold of the metropolis" (Benjamin 104). Serao's women are similarly estranged and alienated, whether they are the objects or the spectators of the narrative exposition. On 1 December 1886, Serao published an article in *Corriere di Roma* titled "La donna réclame," in which she denounced the commodification of women by consumer society. Horrified by the scene of young women promenading in front of Parisian department stores with signs advertising fashionable clothes, Serao demystified the lure of consumer seduction by revealing the alienation produced by public exposure.[19] She was concerned not only about women who carried commercial signs but more generally about the contemporary trend towards cultural and social practices that assigned economic value to women's bodies and turned them into visual sites of consumer desire.

> Ma quante donne senza portare sul mantello il prezzo di costo, servono volontariamente o involontariamente di reclame? Non vi è la brutalità della forma, ma la sostanza resta uguale, dolorosa. Vi è pel mondo un giovanotto nobile come tutti gli altri nobili, ricco come tutti gli altri ricchi; e si addolora che nulla, nulla varrà a renderlo singolare. Una bella donna passa, al suo passaggio tutti si voltano, ella ama il denaro, ella sarà la reclame di colui che la cerca, e il giovanotto sarà conosciuto per lei [...] Oh quante donne inconsciamente o consciamente, sono come le povere commesse che passeggiano per le vie di Parigi, col mantello dal cartellino impresso! E quelle che lo sanno, di essere donne reclame, prendono questa professione con rassegnazione, soffrendo tacitamente, sentendosi offese nella parte più delicata del cuore; e talvolta si rassegnano ma si vendicano. Invece le accecate credono nell'amore, nell'ammirazione di colui che le ama per mostra, per vetrina, per reclame: credono e si sacrificano e si compromettono e agonizzano sublimi di dolore. A un tratto la benda cade, hanno fatto da manichino, hanno fatto da insegna, ciò le getta in un abisso di umiliazione. Qualcuna si perde, qualcuna ne muore. Entrambi ugualmente infelici. L'epoca nostra è triste.[20]

> [How many women, without carrying the price of the cost on their cloak, advertise themselves whether voluntarily or involuntarily? They may not do so explicitly, but the reality remains the same, painful. Take the example of a young noble, like all the other nobles, wealthy like all who are wealthy; and it pains him that nothing, nothing will set him apart. A beautiful woman passes, all turn upon her passing, she loves money, and she will be the trophy for him who seeks her out, and the young man will be known because of her [...]. Oh how many women, knowingly or unknowingly, are like the poor committed women who walk the streets

of Paris, with a cloak and an embossed tag! And those who know they are women advertisements take this job with resignation, suffering tacitly, feeling offended in the most delicate part of the heart; and at times they are resigned, but also take revenge. On the other hand, those who are blind believe in love, in the admiration of him who loves them like an exhibit in a window, as a trophy; they believe and they sacrifice themselves and they compromise and agonize painfully. At some point, the blindfold falls, they have acted as a mannequin, acted as a billboard, and that throws them in an abyss of humiliation. Some are lost, some die from it. Both are equally unhappy. Our time is a sad one.]

It would seem a paradox that Serao criticized the very consumerism that enabled her to flourish as a writer of sentimental fiction. And yet her criticism of the delusions fostered among women by consumerism and romance does not contradict her general approach to sentimentality, which is always anchored in a strong didacticism and, most importantly, never glamorizes love and its effects on women's lives.

If all Serao's sentimental heroines, including the fallen ones, never completely break with female decorum, and so are granted the possibility of redemption, they inevitably find themselves stranded in a state of sterile illusion – similar to that described by Benjamin, where the flânerie succumbs to an inevitable last journey, death – confounded by a dream image that takes the stroller nowhere (105). Serao's love letters, too, can offer only a momentary escape from the confinement and dullness of domestic life: that flight of fantasy, that moment of illusory escape is followed by a repudiation of erotic desire and the return of the strayed women to a virtuous condition that is just as unattainable and fleeting as the erotic desire. Like prodigal daughters, Serao's heroines are always bound to return home. More circular than teleological, more prescriptive than descriptive, Serao's sentimental fiction offers very little scope for a detailed exploration of female erotic desire. Written for women of the middle classes who were expected to see themselves mainly in the role of the wife, mother, or daughter of a respectable man, these stories avoided any explicit reference to or portrayal of erotic desire. The letter format facilitates this fragmented amorous discourse by allowing the author to develop the love story in a sequence of scenes that are structurally independent of one another. In "Falso in scrittura," for example, Serao develops a conventional representation of female adultery by merely alluding to rather than portraying the affair. It opens with the protagonist's attempt to defend her reputation, which has been discredited by a young man's efforts at seduction. As Adriana, the female character, writes in the first of her three letters,

> Egregio signor Cesare – La sua lettera m'ha fatto male. Passai tutta la notte a interrogare la mia coscienza di donna onesta, per sapere che atto, quale parola mia, le abbiano permesso scrivermi quel che m'ha scritto. Ella m'ama, signor Cesare, e vuole essere amato da me – donna vincolata, malamente sì, ma vincolata. E quale idea ha ella della donna, della virtù, dell'onestà? E in quale pessimo ambiente femminile ha vissuto, per creder pessime tutte le donne? [...] Lei sa quali e quante disillusioni ha avuto la mia gioventù. Tutte le mie balde ed azzurre speranze, caddero come foglie inaridite da un albero decrepito. O nella vasta solitudine che ho intorno, non sento voce amica, non vedo uno sguardo luminoso che mi riscaldi. Il mio sogno è scomparso, l'amore del mio cuore è distrutto.[21]
>
> [Dear Signor Cesare – Your letter has hurt me. I spent the entire night interrogating my conscience as an honest woman, trying to figure out which act, which of my words, could permit you to write what you have written to me. You love me, Signor Cesare, and want to be loved by me – a tied woman, unhappily, yes, but still tied to a vow. And what idea do you have, Signor Cesare, of women, of virtue, of honesty? In what terrible feminine environment have you lived, to have such an ill opinion of all women? [...] You know how many disappointments my childhood has withstood. All of my hopes, daring and confident, fell like withered leaves from a decrepit tree. Oh, in the vast solitude that is all around me, I hear no amicable voices, I see no luminous glances that warm me. My dream has disappeared, the love in my heart has been destroyed.]

Serao portrays the female protagonist's initial reluctance to begin an affair with the young man who is pursuing her. She is a married woman and, despite the fact that she is unhappy in her marriage, she invokes the principle of female virtue in order to justify her rejection of the adulterous proposition. "Se vuole, io sarò la sua amica, la sua buona e servizievole amica; lei sarà il mio sincero amico" [If you want, I will be your only friend, your good and helpful friend, you will be my only true friend], Adriana says, although from the following letter we understand that she has indeed betrayed her husband and has fallen into Cesare's arms. A third letter follows, in which the protagonist writes of her decision to stop seeing her lover. It is relevant to note that the second letter, in which the writer describes the adultery, is much shorter than the first and the last. It consists of only a few lines, and rather than describing, it forewarns, as it expresses the futile and disastrous nature of such behaviour. The literary emphasis, thus, falls on the virtuous behaviour described in the first and third letters.

The prodigal daughter *par excellence* of Serao's fiction is Checchina, already mentioned above for the way in which Serao crystallized the image of the nineteenth-century middle-class married woman who temporarily falters in her virtue when faced with the opportunity of experiencing her dream of romantic love. Checchina, bored and humiliated by her joyless marriage, and flattered by the Marquis's gallant attentions, resolves to go to her prospective lover's apartment for a romantic rendezvous. However, as she wanders through the streets of Rome, overcoming the many obstacles that a respectable woman walking by herself is bound to confront, she decides not to meet with the Marquis after all and returns home. What Serao offers her readers here is the image of a threshold – the liminal visual representation of a fantasy world, divided between good and evil, with the female protagonist situated on the brink, on the edge of the abyss of perdition.

While this parable of female virtue (the title, *La virtù di Checchina*, makes it explicit) is not written in epistolary form, it nevertheless provides a background against which all of Serao's sentimental epistolary fiction must be read. Checchina, the stumbling *flâneuse* of Italian literature, with the meanderings of her desire, both at the level of mental momentary escapes from the dreariness of daily life and of physical movement through the public streets, establishes a dialectical relationship with the implied reader, who becomes part of that imaginative world through a set of images that are exemplary in scope and intention. Serao's sentimental epistolary fiction provides readers with a set of moralistic tableaux, as in devotional literature, portraying scenes of women intoxicated by the erotic fantasies made available to them by modern consumer society. Serao's epistolary fiction, thus, combines both erotic and educational narrative purposes, seeking to represent and at the same time didactically reprimand what, by nineteenth-century standards, were considered deviant, excessive female erotic desires. According to Janet Altman, "the content of letter novels divides into two basic categories: erotic and educational." In the first type, the erotic impulse (the desire for the missing lover) generates the epistolary exchange, while in the second type, "the letter writer functions as a guide, and the novel assumes the form of a travelogue or compilation of essays on contemporary society" (196). Serao's sentimental fiction belongs to the latter type.

The letter form facilitates the didacticism of Serao's narrative. The letter is written to instruct, to clarify, and to explain a situation, and in most cases to rectify what the writer of the letter perceives as an incorrect interpretation of the facts. "Signora, voglio scrivere a voi, in quest'ora di morte" [My lady, I wish to write to you in this hour of death], Matilde Serao, wrote in *Gli amanti* (1908): "Solo a voi, signora, mentre non mi conoscete, voglio dire che muoio, uccidendomi, voglio dire perché mi uccido" [Only to you, my lady, while you do not know me, I wish to tell you that I am dying, killing myself, I wish to say why I am killing myself].[22] And in *Lettere d'amore* (1901), the protagonist of the letter writes,

> Mia signora,
>
> Io sono un temerario. Io oso rompere un lungo e alto silenzio: alto silenzio di cui si è sempre circondato l'incomparabile amore, che voi avete acceso nell'anima mia. Ma questo segreto inaudito non può restare chiuso nel mio petto per un giorno, un'ora di più: bisogna che le parole vi dicano, anche pallidamente, quale è stata la passionale mia tortura, dalla sera in cui, senza saperlo, senza volerlo, io vi amai: bisogna che voi conosciate, in questa lettera e da me, tutta la istoria singolare di cui, ohimè, ignoraste sinora il più breve episodio e di cui non potrò che accennarvi tutti i grandi capitoli.[23]

> [My dear lady,
>
> I am daring. I dare break the long and deep silence, the profound love which has surrounded the incomparable love you have sparked in my soul. But this untold secret can no longer remain closed in my chest, not for one day or hour longer. I need words to tell you, however inadequately, what passion and torture I felt when, without knowing, without wanting it, I loved you that night. You must know, by this letter of mine, the whole story which, alas, you ignored in its smallest episodes and of which I will mention only its main chapters.]

A letter is written when the interlocutor is absent, when one is unable to speak in person to the beloved. It is thus written to satisfy the need to send a message. In amorous letter writing, there is always the compulsion to convey a message and to make sure that the message is interpreted correctly. With Serao, the letter is written so that the protagonist can recount the reasons for and the emotions involved in her or his love torments and the recipient can be made aware of them. It is also written so that the reader can voyeuristically observe the dramatic development of an impossible love story and be instructed by it. The letter is used, therefore, as an instrument not only of communication but also of constraint. The reader is forced into a conservative reading of the text that is provided by the writer. Exposed to the grave consequences of illicit love affairs and sinful fantasies, Serao's female readers are taught to read these stories in a way that corrects any possible initial ambiguity. Serao's short stories in the epistolary form conformed to this conservative perception of gender roles and provided her female readership with an exemplary narrative of feminine self-denial. The fact that she wrote these stories using the letter form illustrates her desire to connect with her readers and convey a strong message, facilitating identification with middle-class women (the core of her readership) who looked up to female public figures such as herself, and guiding them through the maze of social conventions attached to a class, the Italian bourgeoisie, that was still taking shape. Strongly influenced by her journalistic work, Serao's sentimental fiction follows many of the conventions of her conduct literature.

Sentiments, and narratives about them, are not necessarily natural expressions of human experience. As Simon Williams suggests, feelings and passions, "whilst central to society and amenable to rational management and control, always threaten to 'overspill' or 'transgress' the socio-cultural boundaries which currently seek to 'contain' them" (146). Serao's fiction of the heart, like sexuality in Foucault's theories, is part of a discursive apparatus that is historically constructed and that seeks to create subjectivity through the containment and management of negative emotions; it is just one perspective among what Foucault defined as a "proliferation of discourses concerned with sex" (*History of Sexuality* 18), a proliferation that in Italy reached its peak during the immediate post-unification period through the publication of newspaper articles, fictional narratives, conduct manuals, and advice literature in general, all focused on defining and prescribing appropriate social and gender-specific behaviour according to the expanding bourgeois ideology. During the latter part of the nineteenth century, a new, "explicit" sexual consciousness arose in the public sphere, as the social, hygienic, and moral behaviour of the individual came under scrutiny and was deemed a relevant indicator of the modernization of Italy.

During this time, love, and by extension sex, began to assume a strong "cultural" value as the affective life – comprising both the physical and the spiritual dimensions – began to be viewed as culturally determined and socially significant, rather than simply the physiological means of human reproduction. Writers, publicists, educators, and politicians, as mentioned in the previous chapter, addressed the various aspects – psychological, physical, legal, artistic, and moral – of what constituted an old and yet evolving concept of love. Within this debate, Serao's sentimental fiction is remarkable for its unique perspective, developed through her close attention to the reality of women's lives and her intellectual response to the works of other writers.

Disillusioned with the idea of love as a romantic, vital, and positive force, Serao embraced a utilitarian notion of it as necessary to the social order, in particular the family, the only place where women, in her view, could find solace and protection from the difficulties and perils presented by life. In contrast to a transcending and idealistic notion of love, such as that developed by the early poets of *amor cortese* or more recently by romantic writers (there could be no Renzo and Lucia in her fiction), Serao's sentimental narratives display the suffering rather than the happiness caused by love. In a conference in Milan on 20 May 1920, remembering Neera, a writer who had died two years earlier and whom she greatly admired, Serao explained the differences between their respective visions of love:

> Ella ha considerato e amato ed esaltato l'amore, solo in quello che esso ha di supremo, come vincolo di anima e di corpi, come fusione di due creature in un'armonia di spirito e di sensi, come terrena compagnia di cuore e di esistenza, come, forse, celeste compagnia oltre la tomba. Neera ha creduto e non ha mai voluto finire di credere a tutta la falsa leggenda dell'amore, quale fu consacrata dai secoli, su migliaia e migliaia di menzogne sentimentali: ella ha creduto alla eternità dell'amore, alla sua universalità, alla sua azione creativa spirituale [...] Questa scrittrice ha chiuso gli occhi della sua mente, per non scorgere l'altro volto della chimera dell'amore, quello della crudeltà [...] chi avrebbe osato innanzi a un così candido apostolo dell'amore nella vita e nell'arte, come era Neera, dirle le semplici e terribili verità, sulla brevità dell'amore, sulla sua caducità, sulla sua fatale fiacchezza?[24]
>
> [She considered, loved, and exalted love, only in its most supreme form, as a connection between soul and body, as a fusion between two creatures in

> spiritual and sensual harmony, as earthly companionship for one's heart and existence, and as, perhaps, celestial companionship after death. Neera believed and never desired to stop believing in the whole false legend of love, consecrated throughout the centuries, on the basis of thousands upon thousands of sentimental lies: she believed in the eternity of love, its universality, and spiritual creative force ... This writer closed the eyes of her mind so she could not see the other side of the illusion of love, that of cruelty ... Who would dare tell that candid apostle of love, Neera, the simple and terrible truths of the brevity of love, its transience and fatal frailty?]

Paradoxically, Serao, whose later career has traditionally been viewed by critics as devoted mainly to sappy romantic love, was actually quite critical of idealized representations of love relationships, fairy-tale–like stories designed to appeal to the female imagination. In spite of her critical position, however, Serao understood the role played by romantic imaginings in the life of the women of her time and incorporated "romance" into her writings, judging that, given love's centrality, it represented an effective, even necessary, thematic tool for a narrative exploration of women's lives and social roles. More than an end – the happy ending to a sentimental story – love constituted for Serao an artistic means that allowed her to articulate her poetics and vision of life. Formulaic and moralistic though Serao's sentimental fiction may seem, it was primarily inspired by the notion that culture had to play a significant role in the process of defining and transforming the social realities of Italy. In this sense, it is useful to keep in mind Thomas Beebee's theory of "functionality," which interprets the epistolary text as a vehicle – in relation to other discursive practices – that enhances the creation of a dialogue within a network of social relationships (169). As elusive as it might seem, this notion of a "network" describes quite well Serao's career-long attempt to create a dialogue with her readers, often with the aid of epistolary exchanges – real or fictional – about issues that mattered to them, as well as to interact with her fellow writers, who, like her, were interested in artistically exploring and elaborating on a topic – love – that presented new and challenging perspectives when observed through the lens of modern life.

Nineteenth-century authors of sentimental narratives were well aware of the role played by the literary text, as well as of their readers' habits, tastes, and expectations, and often fashioned their writings accordingly. Not surprisingly, when describing his relationship to the characters and situations illustrated in his *L'albero della scienza,* De Roberto, for instance, spoke of a "disposizione simpatica" (49), a sympathetic connection, felt by both author and reader, with the topics and characters presented. Such "sympathetic identification" between producer and consumer of the narrative text ultimately promoted the "cultivation of a moral and proper repertoire of feelings, a sensibility" (Hendler 2). In this sense, the sentimental narrative creates a bond between author and reader based on a common sensibility constructed through a shared lexicon of suffering. Such a bond, forged under the influence of the politics of the heart, is perhaps the most obvious characteristic not only of Serao's long and diversified literary career but also of her relationship with her readership. As she proclaimed in 1902 in her literary magazine *La Settimana,* "occorre insomma che un novello, costante, simpatico, tenacissimo vincolo si crei fra gli scrittori d'Italia e i lettori d'Italia per mezzo di questa rivista" [it is necessary, in brief, that a new, constant, sympathetic and strong connection is created between the writers and readers of Italy through this magazine] (De Nunzio Schilardi, *La settimana* 12). As idealistic as the proposition might sound today, this unified body of intellectuals and readers was understood at the time to be the very embodiment of modern Italy, a community that took shape during the Risorgimento period and that, within the context of post-unification Italian culture, constituted a crucial emotional force for the development of a collective Italian identity.[25] When on the first anniversary of *Corriere di Roma,* 25 December 1886, Serao wrote a sort of love letter to her readers – "O amico lettore, sapete che cosa è il giornale? È una lettera d'amore, scritta a voi, chiedente quella purissima corrispondenza d'affetto, che lega lo spirito alla spirito" [Oh, my reader friend, do you know what a newspaper is? A love letter, written to you, requesting that most pure affectionate correspondence which connects one spirit to the other] – and when again, in her novel *Ella non rispose,* she mentioned her "comunione spirituale" (spiritual communion) with her readers, she was referring precisely to that ideal of a shared spiritual and emotional patrimony that many writers in post-unification Italy strove to achieve. Serao's sentimentalism, therefore, and in particular its epistolary production, should be understood as a narrative strategy for the creation of a national identity by way of "imaginative mobility," that is a "sympathetic movement" of identification that would connect the subject to a collective identity (Chandler 25).

In light of this interpretation of sentimentalism, Serao's fiction of the heart emerges as an expression of a socially engaged post-unification culture and is strongly connected to her journalistic work (Tobia 428). In epistolary communication, the sympathetic connection between author and reader – typical of a narrative that conveys a strong message – is all the more apparent. Serao famously conducted an intense correspondence with her readers and published many letters in newspapers, directed both to well-known intellectual figures (as mentioned in the previous chapter) and regular readers. She used correspondence to facilitate communication with her readers – whether in a fictional setting or in an information exchange in a newspaper – and to promote the creation of a dialogue with her readers, encouraging them to respond to her ideas – a notion that contradicts the assumed passivity of the reader of popular literature. "L'opinione di coloro che ci leggono" [the opinion of our readers], Serao wrote, "è spesso più giusta, più vera, più vivace della nostra, perché viene dal centro istesso della vita!" [is often more just, true, and vivacious than ours, because it comes from the very centre of life] (Trotta 187). Serao greatly valued the role her readers played in her efforts as a woman to be acknowledged as an established intellectual figure in a male-dominated literary world. When she began to write her sentimental fiction, Serao was already a best-selling author with a national and international reputation. Most of her fiction appeared in the pages of newspapers that had published the journalism that brought her fame with the reading public. In fashioning her sentimental fiction, Serao relied on that fame, on her established reputation with her readers (often developed through letters). Her long and varied journalistic and literary career spanned an era that she captured brilliantly in what Wanda De Nunzio Schilardi defined as "il suo inventario del mondo" [her inventory of the world], her capacity to describe perceptively and accurately in her fiction the realities, complexities, and contradictions of the lives of women whom she knew or simply imagined (*L'invenzione del reale* 13).

But Serao did not view journalism as a simple chronicle of facts and events. "E' tutta la storia di una società, un giornale, ma è, specialmente tutta la sua vita" [The newspaper tells the whole story of a society, in all of its aspects] (*Il giornale* 29). With these words she passionately defended her profession in a conference at Genoa in 1905. And, comparing journalism to the heroic deeds of medieval knights, she described it as the most valuable instrument in modern times for the regulation of a country's moral and social life. "Mille soldati pronti a vincere e morire, sotto un capo valoroso disprezzante la vita e la morte, sono, infatti una forza di difesa e una forza di attacco: un giornale fermo, tenace, impetuoso e coraggioso sotto la direzione di un uomo che è pronto a tutto perdere per tutto guadagnare, può fare tutto il bene e tutto il male, ed è questo suo potere di distruzione e creazione che soggioga tutte le coscienze avide di energia, di imperio, che avvince e vince tutte le anime assetate di un apostolato per un'idea, per un sentimento" [A thousand soldiers ready to win or die, under a brave leader unconcerned with life or death, are in fact a force of defence and a force of attack: a newspaper that is firm, tenacious, forceful, and courageous, under the direction of a man who is ready to lose all in order to gain everything, can do great good and great evil, and it is this power of his to destroy and create that dominates the conscience of those who want energy, domination, and that conquers and wins all the souls eager to receive an idea, a sentiment] (ibid. 30–1). It is here, perhaps, revealed by her faith in the newspaper as a primary means of connection with the people, that one may find the source of Serao's militancy in defence of a basic argument that writing – no matter the circumstances – was moved first of all by the "impulso di comunicare alla folla le nostre impressioni" [impulse to communicate our impressions to the crowd] (ibid. 37). As she confided to Federigo Verdinois early in her career, Serao cared immensely about having a readership – "Purché mi leggano" [as long as they read me] – and worked hard to obtain one.[26] In light of this priority, both Serao's fiction and her journalism reflected her concern for the communicative function of professional writing, a concern that in artistic terms posed the question of what role intellectuals ought to play in modern society and in practical terms addressed the need to create the sort of modern writing that would be relevant to readers and resonate in their lives.

The question of how to relate to readers had been central in the literary debates of nineteenth-century Italy. At a time when the country was still plagued by high illiteracy rates as well as by linguistic divisions, an aspiring writer necessarily had to confront the question of how to respond to the current historical and literary demands for a more democratic culture, that is, how to produce a literature that was, thematically and linguistically, relevant to those social groups – women and men of the middle and lower-middle classes – who had until then been outside the cultural circulation of ideas. Foscolo, among others, had lamented as early as 1809 that contemporary novels "mancano di quella felicità richiesta in opere destinate a lettori d'ogni classe, d'ogni età e dell'un sesso come dell'altro, una felicità che non è tanto frutto di dottrina e abilità, quanto di uso quotidiano di comporre, osservando le tendenze peculiari del pubblico per il quale si scrive" [lack the appeal necessary for works destined for readers of any class, any age, and both genders, an appeal that is not so much the fruit of doctrine or ability as of a daily practice of composition, observing the peculiar tendencies of the public for whom one writes] (135). But while Foscolo in his desire to make literature relevant to the larger public was still addressing a relatively limited group of readers, Serao, more than half a century later, posed the same question but with the determination to appeal to as wide a readership as possible.

That Serao's relationship with her readers was crucial in her professional career is a fact openly acknowledged by Serao herself. On one occasion, she even polemically suggested that true literary merit originated in the relationship with the public, understood as a necessary agent in the production of literary meaning. When in 1887 she responded to the recurring debate about the lack of an Italian tradition of novel writing (still an echo of Bonghi's polemical pamphlet), she distinguished writers like herself, who always engaged in dialogue with their readers, from those, instead, who pursued artistic creation in isolation, indifferent to readers' needs and interests: "Triste periodo questo in cui i romanzatori o scrivono per il teatro, o fanno giornalismo, o si ritirano in qualche paesello ai solitari piaceri dell'arte che non cerca pubblico! Le romanzatrici sono più coraggiose [...] Le donne che scrivono hanno maggior facoltà e maggior forza di concentramento, non si distraggono facilmente dalle futili circostanze esteriori, il lavoro le assorbe, completamente" [This is a sad period, when writers either write for the theatre, or practise journalism, or retire to some small town where art's quiet enjoyment does not seek an audience. The female writers are braver [...] The women who write have more ability and greater power of concentration, they are not easily distracted by trivial external circumstances; they are completely absorbed in their work].[27] Serao was not only familiar with Bonghi's writings but acknowledged him, along with Francesco De Sanctis, as a major inspiration for her own writing. She discussed this as follows in an article quoted by Neera in *Una giovinezza del secolo XIX*: "O primavera della nostra età, in cui nulla ancora sapevamo esprimere, ma tutto sapevamo comprendere [...] Sapete, allora, come vivevamo a Napoli? In continuo contatto spirituale con Francesco De Sanctis e con Ruggero Bonghi, di cui ogni pensiero e ogni parola erano nostro soave e forte pascolo" [O spring of our age, in which we could express naught and yet we could comprehend all ... Do you know, then, how we lived in Naples? In continuous spiritual contact with Francesco De Sanctis and Ruggero Bonghi, in which every thought and every word were our sweet and strong pasture] (117).

It was perhaps from this Neapolitan intellectual milieu that Serao drew inspiration for her conception of literature as civic engagement and of art as promotion of knowledge. For De Sanctis, in particular, school and literature constituted the main instruments for the creation of a national Italian consciousness. "Solo nello studio delle cose lo spirito esercita ed educa tutte le sue forze, e a questa educazione dee provvedere la scuola" [Only in the study of things does the spirit exercise and educate all its forces, and this education must be provided by the school], he wrote in his *Pensiero educativo* (96), emphasizing the teacher's role as a leader: "il maestro, dicevo io, non dee dogmatizzare, tenersi fuori dell'auditorio, sputar senno e mettere sempre innanzi il suo personcino. Egli dee entrare in comunione intellettuale con la gioventù, e farla sua collaboratrice" [the teacher, I was saying, must not dogmatize, put himself above the audience, spout wisdom and put himself always first] (ibid. 99). And when he praised Pietro Thouar's *Letture di famiglia* and his short stories for children – "ecco un uomo onesto; ecco un uomo che ha cuore. Con che amore lavora in pro della gioventù e con che intelligenza!" [here is an honest man, a man with a heart. With such love he works in favour of youth and with what intelligence!] (ibid. 129) – he was outlining the additional educational role played by literature, a mission that Serao – who was trained in a *scuola normale* to become a teacher – embraced in her popular fiction. In 1855, Bonghi complained about the "poca leggibilità de' libri italiani" [lack of readability of Italian books], and wished that a new literature would soon be born that would be "agile e desta che ritrasse un pensiero comune, se ne nutrisse e lo nutrisse con uno scambio continuo di sentimenti e di eccitazioni tra paese e scrittore" [agile and vivacious and portray a common thought, and nurture itself with it and nurture it with a constant exchange of sentiments and ideas that are common to both country and writer], a literature that, most importantly, "sforzasse i letterati a poco a poco ad uscire dal chiuso dei loro gabinetti, a versarsi in mezzo ad una società cui premesse d'ascoltarli, e dalla quale mettesse conto d'essere ascoltati" [would force literati to slowly come out of their enclosed studies and put themselves in the middle of a society which cared to listen to them, and by which they cared to be listened to] (43). Thirty years later, Serao's popularity among her readers and, above all, her professional approach to writing testified to the changes taking place in Italian letters during a time of great transformation in the way in which the literary text was conceived, marketed, and distributed among readers.[28]

The importance of the professionalization of the writer during this time of unprecedented growth of the print media should not be underestimated. Serao's career and literary production attest not only to the response of intellectuals to the national call to help form a modern Italian cultural identity but also to the way in which the relationship between author and reader began to change around the 1870s. As a journalist, she was forced to respond to the demands of the cultural industry, but as a writer she acknowledged the inescapable connections between art and the mechanism for its diffusion. Serao reacted to Bonghi's complaints that women did not read Italian novels by producing works that resonated deeply with the female public of readers, in accordance with Bonghi's axiom that "se ad una letteratura moderna rimangono estranee le donne, vuol dire che essa non ha vita" [if women remain estranged from a modern literature, it means that it has no life] (67).

Ian Watt notes that, from very early in the rise of the novel, women represented a large section of the growing public of readers, given that they enjoyed more leisure time and led a more sedentary life than their male counterparts (43–5). This assessment may not be as true for Italy, where in the second half of the nineteenth century illiteracy rates were still very high, especially among women; but if numerically there were not many female readers in nineteenth-century Italy, the topic of female education galvanized public debate.[29] Throughout the Ottocento, female education had been a central topic in the debate about the development of a national character, conceived as a set of specific gender, social, and cultural attributes. In the eighteenth century, under the aegis of Enlightenment principles, the issue of universal public and lay education had motivated intellectuals and politicians to address the topic of schooling for women, whose education had traditionally been limited to elementary schooling and devotional instruction that was usually provided by religious institutions. It was initially a question of whether women could, rather than should, be educated: "se le donne siano atte così come gli uomini al governo, alle scienze, e alla guerra" [whether women are apt, as men are, to governing, to the sciences and war], as Antonio Conti wrote in a letter in 1721.[30] The discussion continued in the following century, gaining momentum during the Risorgimento, when the ideal of the family, centred on the notion of a virtuous maternal presence, was established.[31] At this time, the "feminine ideal" ("la pazienza, l'amore, l'obblio e il generoso sacrificio di sé" [patience, love, oblivion and generous self-sacrifice])[32] became a social identity, as women were expected to attend to the needs of others in a spirit of self-sacrifice and self-abnegation. The debate over female education was conducted in accordance with this "feminine ideal." A woman was supposed to be educated not for her own sake but so that she could become a civic-minded and morally irreproachable citizen, able in turn to educate future generations of Italian citizens.[33] In addition to promoting the development of a morally healthy society, educational reforms responded also to the concerns raised during several government investigations in Italy, the *inchieste parlamentari*, which disclosed the poverty and dismal conditions in which many people lived, especially women and children. Women's education was therefore viewed, at least in theoretical terms, as an investment in the country's cultural and economic future – as a way of redressing a problem that, though gender-specific, affected the whole society.

In spite of two important educational reforms (the Casati and Coppino laws), Italian women in the late nineteenth century were still largely undereducated. Nevertheless, though still lagging behind their counterparts in other European countries, they were receiving some education; and with that education they were entering the world of literary production and consumption. Surveys conducted during the second half of the nineteenth century demonstrate that, albeit in small numbers, women were reading. In the Marche and Umbria regions, for instance, between 1869 and 1870 women composed the majority of readers of novels available through membership in *biblioteche popolari* (Porciani 43). And writers like Bonghi, who enthusiastically endorsed the production of Italian novels, identified in women a major constituency as both writers and readers in the future literary market. As Bonghi observed,

> La donna nella letteratura deve entrare più come direttrice che come operaia; allora col suo criterio fine e giusto, con quella sua delicata spontaneità di sentire, con quella sua attitudine a scoprire le piaghe del cuore, con quella sua prontezza ad avvertire le lacune o le parti risibili della natura dell'uomo, con quel suo uso di mondo, con quel suo bisogno di verità e schiettezza, con quel suo vivere nel presente [...] ha una influenza potente ed utile sulla letteratura moderna di un popolo. Oltre di che per il suo posto nella famiglia e nella società è l'istrumento più adatto e più sicuro per la diffusione della coltura, e per la natura delle sue occupazioni potrebbe fornire il maggior numero de' lettori d'un libro. (32)

> [A woman in literature should enter more as a director than as an employee; so with her fine and correct criteria, with her delicate spontaneity in sentiments, with her interest in exploring the troubles of the heart, with her awareness of the frailty of human nature and the ironies of life, with her worldly ways, with her need for truth and frankness, with her way of living in the present [...] [she] has a powerful influence that contributes to the development of a modern national literature. In addition, her place in the family and in society provides a suitable and reliable way to further the dissemination of culture, and by the nature of her occupation she can provide the largest number of readers for a book.]

Women played an important role in nineteenth-century Italian culture, especially in relation to the burgeoning publishing market, not only as educators but also as consumers. Historians and scholars of consumer culture have demonstrated that the growth of a material culture began in Europe well before the actual process of industrialization, and that it was "the desire for consumption more than processes of work that played an active role in giving shape to modernity" (Sassatelli 15). According to this view, modernization was, if not independent of industrialization, at least not exclusively connected to it – a theory that would account for those countries, like Italy, where the Industrial Revolution took place later. While the question of when Italy became an industrialized country is a complex one, the history of the Italian print industry demonstrates that cultural consumerism already existed in nineteenth-century Italy. The large production of novels, newspapers, magazines, manuals, and textbooks that characterized the cultural industry of post-unification Italy occupied a significant portion of what Sassatelli called the "sphere of exchange," the intermediary space between what were considered the two most crucial components of people's existence and daily life: work and leisure, representing, respectively, the public and private domains of life. Furthermore, leisure time in nineteenth-century Italy was not merely a "tempo dell'ozio" [time of idleness], insofar as a strong work ethic pressed people to use their free time in a productive manner (Scarpellini 21). The purchase of a book, therefore, whether for entertainment or edification, not only was a sign of social distinction but also indicated a belief that one should use one's leisure time constructively. Neither exclusively private nor completely public, cultural consumerism was instrumental in harmonizing these two spheres of life and, additionally, in synthesizing what in the nineteenth century came to be considered the complementarity of the two sexes, according to which men belonged to the public sphere of production (work) while women, frivolous and hedonistic, were concerned only with the domestic side of life, including entertainment in all its different forms. It was in the nineteenth century that the notion of woman as the quintessential consumer began to take shape, in accordance with a view that women were more drawn to a type of hedonism that "sees objects as ripe for personal creative fantasy" (Sassatelli 17). According to Colin Campbell, a distinction must be made between modern and ancient hedonism, based on the fact that in modern times consumers desire objects mainly for personal enjoyment by way of day-dreaming. While in ancient times people pursued mainly sensory pleasures connected to eating, drinking, and other forms of social practice, modern individuals gain pleasure not through the object itself but through the symbolic space implied by the object. Under the influence of romantic notions of self-expression, the modern consumer seeks personal fulfilment through the pleasure of the imagination. Hence the relevance of novels, which became a main vehicle for the expression and consumption of such enjoyment, especially for women who, according to traditional as well as more modern sensibilities, were seen as naturally inclined to fantasy, and therefore likely to enjoy the fictional worlds provided by the modern novel (86–7).

If women had a natural inclination towards reading novels, epistolary fiction was viewed as particularly suited to female readers. Because writing letters required a style that was more "natural" and conversational than other forms of writing, women were seen as particularly good at it. But in the nineteenth century, women's capacity to express themselves through writing letters presented a somewhat paradoxical situation. Seen as an example of women's improved education, letter writing was interpreted as proof of women's increasing ability to read and write, but at the same time as a dangerous space within which women could communicate independently and without proper supervision. In this regard, Aristide Gabelli noted in his "L'istruzione e l'educazione in Italia" that "è quasi comune il canone prudenziale che alle donne sia cosa arrischiata insegnare a scrivere, perchè altrimenti se ne servono per fare all'amore" [it is a common, prudent assumption that it is risky to teach women to write, in case they use it to make love].[34] Serao's epistolary fiction is based on the narrative potential of this conflict between expression and repression of illicit female erotic desire. That the author then used women's presumed inclination to fantasize about love to bring them back to the reality of moral and social obligations is not surprising given her belief in art as a cathartic force.

Serao's prescriptive sentimental fiction reflected the author's conservative attitude to the redefinition of gender roles in modern society, but it also resulted from artistic concerns that Serao mulled over during those years. It is not surprising that in the 1890s, when Serao began to write her sentimental fiction, she theorized about the need to infuse literature with a strong spiritual component. In an article published in *Il Corriere del Mattino* on 8 July 1894, Serao defined herself, along with Antonio Fogazzaro and the French writer Paul Bourget, as a "Cavalieri dello spirito," a proponent of a spiritual emphasis new to the contemporary European literary scene. "Ma che è, che sarà questa corrente spirituale nell'arte e nella poesia?" [But what is and will be this spiritual current in art and poetry?], she asked. And if she conceded immediately that as a literary phenomenon that spiritual current was still too recent to have a measurable effect on society (for Serao, always a sine qua non condition of literary production), she defended the inspirations and aspirations that characterized it as a movement:

> Come causa, si può arguire, con facilità, essere un sollevamento dell'anima contro l'aridità, contro l'asprezza di un naturalismo male inteso, contro la vacuità di una verità troppo breve, troppo esclusiva, troppo assoluta. Coloro che hanno creduto instaurata, per sempre, una forma d'arte nel naturalismo, si sono esaltati della loro piccola scoperta e hanno esagerato sino al delirio, non hanno compreso quale ribellione avrebbero causato a quelli che guardano con occhio più quieto la vita e le sue ragioni [...] Un movimento di reazione era, ormai, naturale: e si è sviluppato per fortuna in coscienze intellettuali e sapienti, in anime che sanno leggere in sé stesse prima d'ogni altro, e sanno parlare alla folla. Movimento incomposto, saltuario, bizzarro [...] che importa? Importa che esso sia. Importa che si agiti in fondo al nostro cuore una domanda, un dubbio, una grave incertezza: importa che ognuno di noi si chiegga se tutto quello che ci meravigliò e ci affascinò, in vent'anni, era la verità e non altro che la verità: importa che nel silenzio delle profonde cogitazioni, ognuno di noi ricerchi nuovamente le sorgenti disseccate della sua vita interna e trovi modo di farle ripullulare e, non trovandole, cerchi, cerchi ancora, cerchi sempre; importa che la nostra coscienza non si appaghi, non si cheti, non si addormenti: importa che le ragioni dello spirito ci riappaiono, superiori,

supreme, pacificatrici, consolatrici! Tutta la verità è altrove. Importa di ritrovarla. ("Cavalieri dello spirito" 57–58)

[A cause of this, one might easily deduce, is the soul's mutiny against the aridity, the harshness of bleak naturalism, against the vacuity of truth, too short, exclusive, and absolute. Those who believed that a form of art had established itself forever aggrandized and exaggerated the importance of their little discovery to the point of delirium. They did not understand what rebellion they would cause among those who looked at life and its reasons with calmer eyes. A movement of reaction was, by then, natural; and it developed luckily in wise and intellectual consciences, in souls which can anticipate the reactions of others and know how to speak to the crowd. A movement that is incomplete, fitful, and eccentric – but it doesn't matter. What matters is that it exists. It matters that a question, a doubt, a strong uncertainty agitates our heart. It matters that each of us asks whether all that fascinated and amazed us in the last twenty years was truth and nothing but the truth. It matters that in the silence of our profound thoughts, each of us searches anew for the dried-out springs of our internal life and finds the way to replenish them, and, not finding them, searches, searches again, always searches. It matters that our conscience is not easily gratified, is not silenced, is not fallen asleep. It matters that our spiritual impulses reawaken, superior, supreme, pacifying, and consoling. Truth is elusive. It is important to find it again.]

Reading Fogazzaro's response to Serao's letter, it becomes apparent that there was no real agreement on what the "Cavalieri dello spirito" ought to represent as a movement.[35] It never became the new literary current that Serao hoped to see as a competing force to the still predominant narrative school of naturalism. And yet Serao's self-proclaimed sense of spiritual knighthood constituted an important statement about how she conceived her work in close relationship with other authors and through an envisioned network of spiritual connections with her vast public of readers.

As opposed to naturalism, this spiritual focus was for Serao a "viaggio di esplorazione, di scoperta" [a voyage of exploration, discovery] through life experiences in pursuit of artistic truth, a focus that she saw as a tool of narrative experimentation and ideological investigation ("Cavalieri dello spirito" 58). Imbued with idealism, which for Serao had for too long been negated by naturalism, her new self-defined spiritual approach to art strengthened the thematic focus of her writings of those years on the exploration of the emotions of the human heart.

Among all human emotions, love reigned supreme for Serao, as Angelica affirms in "L'ultima lettera": "perché l'anima vuole il suo pascolo, perché è l'amore che serve all'anima, perché niuno sfuggirà a questa necessità dell'anima" [our soul wants its own pasture, because our soul needs love, because no one will escape this necessity of our soul] (*Gli amanti* 289). But central to the human soul though it was, love remained for Serao an unattainable goal. There was no happy ending in Serao's sentimental fiction, which, paradoxically, highlighted the fleeting and illusory nature of any vision of ever-lasting happiness through love. This was made clear in several of her short stories. In "Lettera di amore," included in *Novelle sentimentali,* the protagonist, a man this time, sends his prospective wife a "love letter" in which he tells her of the disappointment he experienced in meeting her in person for the first time after three years of long-distance correspondence that had given him a false picture of her and false expectations about their relationship. Seemingly cruel, this letter was presented as a true act of love because it was written, as the protagonist declared, "per strappare [...] la benda dagli occhi, per indicarvi la verità" [to take off the blindfold from our eyes in order to show the truth] (15). Far from being the love letter that a person fantasizing about love would wish to read, this letter epitomized Serao's general approach to sentimental fiction, which debunked rather than reaffirmed the fairy tale nature of conventional sentimental fiction. In another collection, *Lettere d'amore,* significantly subtitled "Il perché della morte" [Why death?], Serao's unromantic approach to love was even more evident. "La statistica sentimentale e sensuale che interroga i cuori e i nervi umani, ahimè, parla chiaro e forte" [The sentimental and sensual facts, interrogating human hearts and nerves, alas, speak loud and clear] declares Clara, the protagonist of letter five when she decides to leave her lover because "non voglio lasciar fare al destino, il quale assegna un limite di tempo quasi matematico alla passione" [I don't want to leave it to destiny, which assigns an almost mathematical time limit to passion] (116):

> Da uno a tre mesi, fantasticheria amorosa; da tre mesi a sei, capriccio violento; da sei mesi a un anno, amore. Oltre un anno, passione: e purtroppo, non oltrepassa il limite dell'anno secondo, compiuto, per la fine della passione, nella sazietà e nella stanchezza. (110–11)
>
> [From one to three months, a love of fantasy; from three to six months, violent caprice; from six months to a year, love. Beyond one year, passion; and unfortunately it does not last beyond the end of the second year, marked by the end of passion, by satiety and fatigue.]

Serao felt almost an obligation to reveal her pessimistic view of love. In *Sognando* she explains that the artist is obliged to expose love's fallacy:

> E giacché nessun psicologo, ahimè, nessun artista ha potuto negare la tremenda verità ed è che l'amore sia un sentimento breve e fallace [...] giacché le limitazioni, le imperfezioni, le delusioni dell'amore non le può negare nessuno, giacché esso sovra tutto è breve, breve, breve. (94)
>
> [No psychologist, alas, no artist can negate this terrible truth, which is that love is a brief and false sentiment ... since the limitations, imperfections, and disillusionments of love cannot be negated by anyone, since it is most of all brief, brief, brief.]

And to clarify any possible confusion about what love truly is, she reminds her readers that "l'ascensione al vertice è lenta, si dice: ma si discende precipitosamente" [the rise to the apex is slow, they say, but the descent can be precipitous] (111), stressing the danger implicit in such a fall. Finally, in *La vita è così lunga*, composed of two sections (the second significantly subtitled "L'amore è così breve"), Serao describes her artistic interest in "il cuore inquieto" [the restless heart]:

> Io sono felice; anche voi siete felice, malgrado me lo abbiate detto. Cerchiamo di essere meno felici, desiderando qualche cosa di bello e d'impossibile ... Il desiderio è, in sé, un delicato piacere. I bambini sono pieni di desideri, vogliono questo, vogliono quello, e finiscono per desiderare la luna. Quando filano le stelle il popolo dice che bisogna formare subito un desiderio, il quale sarà esaudito. Siete orgogliosa e arida e non amate punto i bambi, ma non negherete che l'infanzia e il popolo fanno spesso della poesia non rimata. Quando avrò trent'anni e sarò tanto vecchio, scriverò un trattato, un libro di scienza sul desiderio. (143)
>
> [I am happy; you are also happy, even though you had to tell me so. Let us try to be less happy, desiring something beautiful and impossible ... The desire is, in itself, a delicate pleasure. The children are full of desires, they want this and they want that, and end up wishing for the moon. When the stars fall, people say we must make a wish, which will be granted. You are proud, distant, and do not love children, but you cannot deny that the youth and the people often produce unrhymed poetry. When I turn thirty and am very old, I will write a treatise, a scientific book on desire.]

Because Serao was a writer and not a scientist, she wrote fiction, in particular fiction focused on female desire. "Tu, al solito, dirai che io studio troppo le donne e poco gli uomini: ma lo scibile è troppo vasto [...] e solo per il problema femminile non basta la vita di un uomo" [As usual, you will say that I study women too much and men too little, but knowledge is too vast ... and to understand women's problems alone, one human life is not long enough] (*La vita è così lunga* 20–1). Serao defended herself against the charge that she created fiction that focused too exclusively on women, and she was unapologetic about her sentimental fiction, which she viewed as a legitimate artistic exploration of a central aspect of women's lives. With her fiction of impossible loves Matilde Serao advocated scepticism towards the fairy-tale dream of eternal love, while pursuing the artistic exploration of women's inner world: their dreams, desires, hopes, and frustrations. As she stated in an article published in 1885, "Se l'arte deve intendere a descrivere la lotta dello spirito umano che trionfa delle cose e di sé stesso, se veramente tutto quello che è, si muove, si agita, combatte, soccombe o vince, è dominio dell'arte, perché volete voi negare a tanta parte del mondo donnesco i suoi poemi, le sue storie, i romanzi della sua crudele battaglia?" [If art must aim at describing the human spiritual fight that wins over things and oneself, if really all that is, moves, agitates, fights, loses or wins is the realm of art, then why do you want to negate so much of the female world – the poems, stories, novels about this cruel battle?].[36] Writing her sentimental fiction was Serao's way of recognizing women's inner struggles, and though she was pessimistic about women's lot, often comparing it to a "cruel battle," she never considered it not worthy of artistic representation. On the contrary, a woman's struggle was for Serao a key source of inspiration for the many stories and novels she wrote over the course of her literary career.

Writing about love and the human heart was Serao's way of expressing her unique vision of the creative process, the moral purpose of literature, and her own identity as a post-unification author of Italian literature. As she stated in Paris in 1899 in an interview with Rowland Strong, the *New York Times* correspondent, who described her as the female Italian counterpart of D'Annunzio, "I think that a novel to be successful ought to appeal in some measure to the moral sense. I cannot quite agree with those who worship only beauty of form. I am a very great admirer of D'Annunzio, who is our prophet of beauty in Italy, but I fancy that in late years he has lost touch with the human heart; he has become obscure because he is not sufficiently human."[37]

5 Conclusion

The nationalization of the postal service marked a historical moment of transformation for the post-unification development of letter writing. Epistolary fiction became part of this cultural, institutional, and technological process as well as playing a role in the formation of modern national identities. The postal service came to be perceived by contemporaries as an indicator of progress and a provider of a wider and more efficient circulation of information, and readers and writers of letters relied on a shared set of conventions and perceptions that transformed the letter into a bridge between the private world of personal communication and the public arena of information exchange and expression of public opinion. Whether in private correspondence or in a public forum, letter writing acquired the significance of a social event, as epistolary communication required correspondents to inhabit a space other than the one they would normally live in, one in which new information could be gained. At a time when most people in Italy still lived in rural areas, far from urban cultural centres, letter writing allowed readers to travel distances, actual and imagined, and share experiences observed from the comfort of their domestic realm. Literature capitalized on this imaginative space and enticed readers into a fictional realm that drew on but also shaped the world the readers lived in. In epistolary writing, the imaginative space is symbolically located in the letter, the narrative place where writer and reader are united in a process of communication that necessarily enhances a plurality of presence; as Janet Altman reminds us, "the epistolary experience, as distinguished from the autobiographical, is a reciprocal one" (*Epistolarity* 88). Because of this process, letter writing in post-unification Italy fostered a collective awareness, a sense of belonging to something or someone other than oneself. As a growing number of people read and wrote letters, they began to see themselves as part of a national community that viewed the letter as both a channel in the process of information exchange and a necessary instrument for the education and modernization of the nation. Ordinary citizens wrote letters to the editor to express their opinions, intellectuals and journalists used the letter to create public opinion, and writers of novels and short stories used it to enhance both the formation of a national public of readers and their own standing as national literary figures.

If it is true that nineteenth-century Italy did not produce great works of literature in the epistolary form, my research nevertheless demonstrates that some of the main literary figures of the time were practitioners of epistolary writing. A focus on the use of the letter during this period provides an exceptionally "functional" tool, a theoretical framework but also an actual body of Italian texts through which to explore a literary period that is often neglected because of the heterogeneous nature of many of its productions. Several studies have recently been published on the Italian epistolary tradition (see, for instance, Gino Tellini, Gabriella Zarri, Adriana Chemello, Diana Robin, Meredith Ray, and Laura Salsini, among others), but they tend to focus either on correspondence or on epistolary fiction, without addressing how the different practices of letter writing overlapped with each other. In contrast, my approach has been a synchronic one, an attempt not only to understand better what has been defined as the "envelope of contingency" of a literary text – the historical, social, and cultural environment in which a text is embedded – but also to connect the higher and lower forms of cultural production in the literature of the period, which, though imperfect and hybrid, still marked a crucial moment in the formation of Italy's modern literary sensibility. Starting, therefore, from the premise that the development of a "postal culture" – modern communicative practices revolving around the letter – stimulated a wider circulation of ideas and information, the present study has explored the different types of epistolary communication – in periodicals, epistolary manuals, correspondence, and fiction – and found that they shared thematic, stylistic, and linguistic choices as well as a readership that viewed all these types of texts with similar expectations of entertainment and instruction. Different as all these texts may seem, they all exemplify a fundamental awareness that letter writing in the post-unification period implied the use of a system of communication unavailable a few decades earlier and dependent on a set of interpretive tools acquired through education. In practical terms, this meant that the authors of epistolary novels, epistolary manuals, and newspaper articles written in the form of a letter adopted a language and style that paid less attention to literary or academic conventions than to the way people communicated in their own correspondence; and readers showed their appreciation of what they recognized as an authentic representation of reality by purchasing these texts and, in the space devoted to the letters to the editor, by participating in the debates of the time. The authenticity that traditionally characterized epistolary writing gained renewed importance in the post-unification era, for it epitomized the urgency of finding a national literary tradition that reflected the realities of the times. In their epistolary fiction, both Verga and Serao attempted if not to represent, certainly to experiment with, the notion of authenticity; and their use of the letter as a literary form attests to their experimentation with communicative strategies.

As letters circulated more widely, they helped produce a national network of social and cultural identities. Cultural practices thrived thanks to this communicative agent, and sentimental fiction, as part of the larger production of a popular fiction rhetorically steeped in emotion, fostered a sense of belonging to a collective, national identity. It may be said that the letter became a carrier of a post-unification cultural message of shared national sentiments. As recent studies on emotions have shown, feelings represent a valid response to specific circumstances and can strategically contribute to the creation of textual empathy and to "the emotional resonance of fiction, its success in the marketplace, and its character-improving reputation" (Keen vii). Verga's and Serao's sentimental fiction relied on such an empathetic response and sought to engage the growing public of readers on issues such as the family, individual versus collective identities, gender relations, women's education, and, more generally, social and gender roles in a fast-changing society – all topics deemed relevant to the social and cultural makeup of what was envisioned as a modern Italian nation.

What is usually referred to as popular fiction, the vast corpus of late-nineteenth-century novels and short stories infused with educational purpose and written for the burgeoning public of newly educated men and women – a corpus to which both *Storia di una capinera* and Serao's sentimental fiction belonged – was, as this study aims to illustrate, a genre adopted by authors from both sides of the gender aisle. The seemingly gender-divided world of nineteenth-century Italian culture (with women authors writing for female readers according to a specific female experience and voice) was, when looked at through the lens of popular fiction, relatively gender inclusive. I am not suggesting that nineteenth-century women writers did not claim a style and a voice of their own, or that they were not aware of or even relying on the fact that their audience was mainly composed of women; rather it seems that their voices were part of a larger choir, a vast and disparate cultural world inhabited by both male and female authors, who, pursuing their own writing careers, expressed similar artistic, ideological, and professional concerns. If it is true that women enjoyed an audience composed mainly of women, it is also true that both male and female authors wrote popular fiction and for a shared audience. A growing number of middle-class women entered the world of cultural consumerism and read the authors and novels that reflected their cultural tastes and expectations. Verga's and Serao's epistolary productions enjoyed a similar reception among female readers (Treves's comment on women in Milan being enthralled by *Storia di una capinera* testifies to this [see chapter 3, note 2]); and both authors expressed similar artistic concerns about the most appropriate language, style, and literary form to use to explore their chosen themes. Despite the differences between Verga and Serao, their epistolary fiction epitomized post-unification cultural efforts to reflect the remarkable transformation characterizing the literary life of a country recently embarked on a modernizing process – a scene marked by great hopes and expectations, but also by deep contradictions and confusion.

Verga's and Serao's sentimental epistolary fiction, traditionally viewed as their "minor" fiction, instantiated the ambiguities and contradictions not only of their own poetics but of the initial phase of development of the modern Italian novel. As Verga attempted to make a national debut with his epistolary novel, he resorted to a literary form – the letter – that on the one hand accommodated a narrative of personal suffering, recounted in confidence by the protagonist to her friend, and on the other enticed readers, through specific thematic, linguistic, and stylistic choices, to see in Maria's story what I would describe as a narrative of enunciation, the formulation of a question about the role of the family and the place of the individual in the smaller and larger community that reflected not only Maria's experiences but those of readers themselves. Similarly, Serao's fiction, far from being merely escapist, engaged those same readers by providing them with a fiction that spoke about their own concerns, hopes, and disappointments in life. Verga's and Serao's sentimental fiction, though artistically not fully original, nevertheless constituted a meaningful attempt to bridge the gap between life and art. A literary critic has recently lamented that a fundamental "divorzio tra scrittori e popolo" [divorce between writers and people] and "incapacità della tradizione di rendere conto del mondo" [incapacity of the tradition to reflect the world] characterized the history of Italian literature at least until the 1950s, after which writers began to break with literary tradition and search for modern and effective forms of narrative expression that reflected "lo schiudersi di nuovi stili di vita e l'attenzione a riferimenti culturali legati, non al mondo delle lettere, ma appunto dell'intrattenimento, della tecnologia e del digitale" [the appearance of new lifestyles and the attention to cultural referents connected not to the world of letters but to that of entertainment, technology, and digital information] (Brevini 70, 151). We ought to ask, however, whether the epistolary fiction that Verga and Serao produced did not in fact reveal, in its experimentation with language and its focus on the letter as a modern means of communication, a glimpse of that very modernity that Italy fantasized about in the post-unification period and achieved only about a hundred years later. In other words, if the sentimental writings of Verga and Serao were indeed intended to be a literary contribution to solving the seemingly insoluble problem of the failure of Italian literature to appeal to a wide readership, as Bonghi had noted earlier, then we might want to reconsider the definition of "minor" traditionally attributed to them and regard them not as aesthetically flawed but as successful in appealing to the reading public – as not "divorced" (to use Brevini's expression) from the realities of the people. If, finally, there is any concluding argument to this research, it is one inspired by the realization that not enough scholarly attention has been devoted to the popular fiction of post-unification Italy and that a new map of the literature produced and consumed during this period ought to be drawn. A literary history that ignores the works actually read by a significant number of readers is no longer tenable. Because of Italy's high post-unification illiteracy rates, it is often assumed that Italians did not read much. Studies of popular fiction as well as of the history of publishing paint a different picture. Italians did read and, whether they read foreign novels in translation or works by their favourite authors (Serao, De Amicis, Fogazzaro, and Verga, among others), they constituted a modern public of readers. Further investigation is called for into the relationship between authors and readers in post-unification Italy, a relationship that undoubtedly influenced the growth of modern Italian cultural sensibilities and identities.

APPENDIX

Letters Transcribed from Newspapers

La Ricamatrice,[1] XII. 6 (16 marzo 1859)

CORRISPONDENZE

Sulla convenienza per le fanciulle di adoperare anche nell'uso domestico la lingua comune italiana[2]

Alla signora Teresa B ... M ... a Mantova

Voi ne eccitate, colta e gentile Teresina, a persuadere le nostri giovani campaesane della necessità di abbandonare anche nei colloqui famigliari l'usanza del rozzo e vario dialetto per quella più nobile e sicura della lingua comune. Oltreché, questa è quistione di gusto e di squisitezza di costumi, vi s'intromette anche un interesse diretto di civiltà. Voi avete non una, ma mille ragioni; ed è per questo che nel volervi accontentare, siamo piuttosto imbarazzati dalla quantità degli argomenti che dal loro difetto.

Poche lingue moderne, saremmo anzi per dire nessuna, vantano maggiori varietà o gradazione della nostra. È un immenso serbatoio di materia parlata, donde la forza dell'uso e il criterio degli scrittori, estraggono senza saperlo di secolo in secolo il nerbo più vivo della lingua comune; e questo si osserva che quel periodo fu più virile e fecondo della letteratura nazionale, anzi della storia nazionale, quando fu più attivo questo lavoro di trasformazione. Del resto, il tempo delle letterature provinciali è passato, né si farebbe ora buon viso a chi traducesse il Tasso in Bergamasco o l'Eneide in Napoletano; si direbbero capricci di begli umori e nulla più. Il dialetto è opportunissimo in chi attende a sminuzzare i primissimi elementi scientifici al volgo che sa leggere, ma non sa ancora comprendere, ma per chi ha raggiunto quel grado di coltura che consente l'intelligenza di qualunque idea anche espressa nella nostra lingua comune, consiglieremo di trascurarlo affatto.

Una fanciulla che esce di collegio colla bella lingua sulle labbra, dovrà ella rinnegarla per adottare il gergo delle cameriere, o non piuttosto dovrà giovarsene sempre, per far partecipe la gente che la circonda del vantaggio guadagnato mediante parecchi anni di istruzioni o di buone abitudini? Nel primo caso è un valore che si perde, mentre s'aumenta, si moltiplica indefinitamente nel secondo; e tanto basta per segnare la via della convenienza e del dovere.

Non diremo che vi si abbia a insistere pedantescamente; ché la pedanteria è la ruggine di ogni buona costumanza, ma soggiungiamo che, s'anco si dovrà ripetere un commento, una parafrasi in dialetto, non sarà male che la prima esposizione sia in lingua purgata.

Sarà anche qualche caso in cui ad una donzella parrà sconvenevole l'adoperare frasi tornite e grammaticali, mentre tutta la brigata non sa esprimersi che in dialetto; ma mancano altre maniere per dimostrare che quella bella diversità non procede né da sussiego né da leziosaggine? La modestia degli atti, l'inflessione della voce, la parsimonia delle parole varranno a ciò assai meglio d'una sciocca condiscendenza all'usanza peggiore dei più. Crediamo che, adoperata a questo modo, la lingua comune farebbe vergognar gli altri della loro scarsa coltura, e gli inviterebbe all'emulazione, o almeno costringerebbe il loro vernacolo a ingentilirsi, a raccostarsi a quel tipo perfetto della bellezza materna: piccolo e primo passo ad utili maggiori.

Di più si sa quanto sia facile alla frase vernacola il trascorrere nelle trivialità, e quale sforzo parrà soverchio ad una donzella bennata per sottrarsi da tale pericolo? Avvezza ch'ella sia a parlar sempre in pretto italiano, non sarà costretta a balbettare e a mendicar le parole quando si trovi in una adunanza che ha il buon senso di non favellare altrimenti; e le frasi le occoreranno pronte e naturali alla penna quando le sia mestieri esprimersi colla scrittura, occasioni che oggimai per fortuna non iscarseggiano neppur al sesso gentile.

Si discorreva or è qualche tempo della convenienza di stabilire in Toscana un istituto di aje, appunto per ispanderle poi in tutta Italia e diffondere così l'abitudine del pretto parlare; finché non abbia effetto un così bello ed utile divisamento, assumano le nostre zitelle, le nostre spose, le nostre giovani madri l'ufficio di quelle aje; mantengano, se le hanno, la buona usanza di parlar bene ed italiano, si sforzino ad apprenderla se non l'hanno ancora, e la gente di casa, i figliuoli, le sorelline, le cameriere sieno obbligate poco per volta a comprendere ed a parlare quella nobile lingua che fu strumento di civiltà al mondo, e sarà sempre pel nostro paese una gloria, e più che un segno una parte non minima di riconoscimento e di vita. Avranno portato il loro sassolino al grande edifizio del rinnovamento civile; né sarà poco.

Queste brevi parole abbiatele per indizio di premura a compiacervi, voi, o gentile Teresina, che tanto caldamente amate la lingua nostra, quanto squisitamente la parlate e scrivete.

Tutto vostro

N.N.[3]

La Ricamatrice, X. 4 (16 febbraio 1857)

La donna italiana considerata in riguardo all'educazione civile e sociale

LETTERA PRIMA

A Caterina Percoto,
Che cosa direte, buona Caterina, quando mi vedrete prendere la via di Milano ed il mezzo d'un giornale, per venire a conversare alquanto con voi nel vostro campestre soggiorno? Ammetterete per buona la mia scusa, che quella è la più breve per me in questo momento, e che seguo l'uso ormai radicato di pensare in pubblico come un giornalista? Se non date la passata alle mie scuse, ve ne addurrò un'altra delle ragioni, quella del *poscritto*, ch'è la vera.

Io volevo scrivere qualcosa sul tema, che sta a titolo di queste lettere che v'indirizzo. Ma confesso, che quando volli mettermi a scrivere alcuni pensieri, cui avevo per alcun tempo rimescolati nella mente, provai, nella mia qualità di uomo, dell'imbarazzo, e direi quasi, se non temessi di pronunciare una frase ridicola, un certo pudore a parlare dell'educazione delle donne. Rammentai una risposta da voi datami altra volta, ed in cui vidi congiunta una rara finezza e coscienza d'autrice. Dopo letta una delle vostre novelle, in cui sì felicemente s'appajano lo spirito di osservazione coll'affetto, la poesia colla semplicità, richiesi perché a un carattere d'uomo che vi figurava, e su cui eravate passata con qualche leggiero tocco, non avevate dato tutto lo sviluppo a cui il racconto prestavasi. La risposta fu nella sua modestia sapiente. Mi diceste, che voi donna sentivate col cuore di donna, e non vi trovavate ben sicure di penetrate addentro e d'intendere quello dell'uomo.

Ed io, memore di quella lezione, avrei dovuto rinunciare a discorrere di donne alle donne, per il ben giusto timore di parlare di quello che non pienamente conoscessi ed intendessi. Difatti rinuncio, non a sentire e ad ammirare ciò che di delicato e di alto vi può essere ed è spesso un cuore di donna, ma bensì a descrivere quello che sento ed ammiro. Però, sotto a un certo aspetto, sarà lecito anche a me il parlare della donna alle donne; sotto a quello che vogliamo attenderci dalle donne nostre per la civile e sociale educazione nel nostro paese. In tal caso, potranno le lettrici prendere la parte che loro sta bene, lasciando il resto, e mettendo a carico dell'uomo tutto quello in cui non c'intendessimo. Ma anche per questo, o Caterina, ho bisogno che voi donna m'introduciate nella società del vostro sesso, cattivandomi presso le altre un po' di quella benevolenza che mi accordate. Ed eccovi tutto il segreto della lettera, che, per venire a voi, a poca distanza da Udine, prende la via di Milano e si mette nelle colonne della *Ricamatrice.*

Per entrare in materia al modo che si usa, io dovrei spifferare di botto una dissertazioncella sull'*emancipazione della donna*. Ma in coscienza non lo posso fare, e per molte ragioni. Emancipare la donna! Soggetto meschino per uno che vorrebbe emancipare tutto, e che, in fatto di emancipazioni, dalle donne s'attende più ajuti ch'esse medesime non ne abbisognino. Poi, un bel guadagno sarebbe il mettersi in riga con tanti altri emancipatori! Se bene io li guarda, vedo nel maggior numero di costoro, quando non sieno stucchevoli pedanti che ripetono grandi parole, che hanno un povero significato; vedo, dico, degli adulatori e bene spesso dei tiranni della donna, o, come direbbero i nostri vicini, *des exploituers.*

Vogliono emancipare la donna! Ma da che, e da chi? Forse da quelle doti e prerogative che la fanno essere donna, e come tale diversa dall'uomo, da' suoi doveri nella famiglia e nella società, dagli uomini che le donne amano, stimano, e che in esse s'inspirano e si compiono? Vogliono, per emancipare la donna, farne qualcosa di simile all'uomo, la sua scimmia, nella guisa appunto che molti uomini della nostra società credono di emanciparsi col rendersi simili alle donne? Quello che Iddio congiunse l'uomo non separi: e perché le donne e gli uomini sieno congiunti, occorre soprattutto che le donne rimangano donne e gli uomini uomini , che la bellezza, la grazia, e la virtù delle une sia un'attrattiva continua alla forza, alla generosità degli altri, e viceversa. I nostri dozzinali emancipatori non pensano e non intendono, che si devono emancipare ad un tempo e uomini e donne; e che per questo bisogna appunto studiare come gli uni e le altre possano rimanere quel che devono essere e perfezionarsi secondo la loro natura ed in armonia alla società intera.

Se s'intende di dire, che molti preguidizii sociali, molte ingiustizie anche pesino tuttodì su alcune, su molte donne, io sono d'accordo con loro. Ma ci appelliamo a voi, donne nostre, siamo noi propriamente i tiranni che dicono, come uomini e perché uomini? Soffriamo noi meno di voi dei pregiudizii e dei difetti sociali? Non riconosciamo noi anzi la legittimità del vostro impero? Non v'invitiamo fino talora ad abusarne, od adulandovi a guisa di cortigiani che non stimano né la propria né l'altrui dignità, o mostrandoci al vostro cospetto minori di noi medesimi? No, che la donna, la donna della colta società non ha da chiedere la sua emancipazione, ma piuttosto da comandare; certa che sarà obbedita, se comanda prima di tutto agli uomini di essere uomini e come uomini stimabili, e se essa sa essere donna ed adempierà gli ufficii nobilissimi della donna in tutte le età e condizioni. Ed è questo a cui si dovrebbe mirare, illuminandosi e confortandosi a vicenda.

Le donne della società brillante possono anche oggidì, come quelle del medio evo, tenere le loro corti d'amore e comandare a molti cavalieri le ardite e nobili imprese, quale pegno del loro affetto, della loro stima, quale indizio dell'omaggio che vogliono prestare alla bellezza. Schiave non sono, se non le donne che s'incantenano a passioni, a desiderj, a costumi indegni di loro; ma quelle in cui alla bellezza s'associano l'amabilità, la virtù, il sapere, imperano purché lo vogliano. La donna letterata poi, anziché sdottoreggiare di emancipazioni d'altro genere, di emancipazioni che sarebbero la peggiore delle schiavitù, perché tendono a corrompere, avrà bene di che occuparsi nell'educazione e nel miglioramento dello stato delle donne delle classi inferiori. Si faccia pure avvocata del suo sesso, ne parli dei diritti di quelle donne, a cui nella società è serbata una trista sorte, e che non partecipano in eguale misura ai beni comuni, che soffrono ingiustamente, che danneggiano l'umano consorzio col non essere mantenute al loro luogo e rispettate nella loro dignità: ma non faccia delle proprie passioni scusa ad emanciparsi dai suoi doveri, a sottrarsi a que' vincoli salutari che non legano le sole donne, ma e le donne e gli uomini ad un tempo tengono connessi al corpo sociale, gl'infondono vitalità. Tuoni pure essa contro il mercato d'anime che si fa talora, accoppiando, come faceva il tiranno Mesenzio,[4] esseri viventi a cadaveri, gentili donzelle piene di vita e vecchi infrolliti in ogni putridume. Faccia conoscere quanto spesso l'educazione della donna sia falsata e con quanto danno della società. Mostri come nuocano alle felici unioni, ai bene assortiti connubj e pregiudizj di casta. Questi saranno altrettanti modi di contribuire all'emancipazione di tutti e due i sessi: ma ci si faccia grazia una volta di non più ripetere una parola tanto abusata, e che massimamente dalle donne non potrebbe ormai usarsi, senza destare il sospetto di aspirare alla *bloomery*[5], d'intendere, cioè, che la donna abbia a farsi simile agli uomini, snaturandosi e perdendo le sue virtù senza acquistare le nostre.

Mi sembra di udirmi chiedere il tipo ideale della donna, quale intenderei ch'esser dovesse. Io non domanderò questo tipo alla poesia, non lo cercherò fra le Muse. Il genio ispiratore non si sottopone a regole. Ci giova prenderlo e seguirlo dovunque si manifesta quale alito della divinità sopra il mondo umano. Procurerò di formarmi il mio ideale di donna come donna, e fra le donne cercherò quello della donna italiana, e soprattutto della donna che si conviene ai nostri tempi, per il rinnovamento civile e sociale, a cui essa per la sua parte deve contribuire.

Prima di tutto però devo allontanare, se mai fosse entrato nella mente delle mie lettrici, il sospetto, che rigettando io quale regola comune il tipo dell'Aspasia, non intenda di confinare, come altri, la donna alla casa ed alla cucina dandole l'ufficio di economa soltanto, come la donna dai Proverbj lodata perche sapeva tenere in ordine la famiglia, e sull'altro. Avendo di soggiungere qualcosa su ciò, riserbo a parlarne in altra lettera.

Pacifico Valussi[6]

LETTERA SECONDA

A Caterina Percoto,

Con buona pace di molti valenti letterati che pajono di quest'opinione, io non posso o Caterina, trovare un tipo degno d'imitazione per la donna italiana dei vostri giorni nell'Aspasia, cui vollero fare ispiratrice della coltura del secolo di Pericle, perché alle sue conversazioni bazzicavano i dotti della Republica d'Atene; ma per questo, vi dissi, non intendo né che le donne abbiano ad essere (come fra quei bravi Turchi della di cui sorte ed indipendenza tanto s'occupo il mondo incivilito, e fino a dimenticare più prossimi doveri) soltanto uno strumento dei piaceri dell'uomo, né le serve di casa, come taluno vorrebbe farle. Per certuni la donna non ha bisogno di saper molto, né di partecipare largamente ai beni dell'intelligenza. Sappia occuparsi de'suoi lavori femminili, attendere alla biancheria ed al bucato, sorvegliare la cucina e dare la giusta dose a pasticcetti, fare l'economia di casa e badare che per i proprii danari s'abbia tutti i suoi comodi; ed allora essa è la donna saggia e prudente , la donna descritta nei Proverbi da Salomone, il quale, d'altre parte, per i suoi minuti piaceri, si trattava in donne quanto qualunque sultano. Che in certe condizioni sociali la donna non possa aspirare una coltura superiore, lo trovo ben naturale, poiché questa è la sorte del più gran numero anche degli uomini; ma trovo d'altra parte questo giusto, altrettanto utile alla società intera, che la donna, in tutto quello ch'è conforme alla natura sua ed agli ufficii a lei particolarmente riservati, sia istruita e colta, e partecipi nella più larga misura possibile ai beni dell'intelletto, cioè a quanto di più alto e di piu degno può aspirare l'uomo sulla terra. Anzi, se considerassi la donna sotto l'aspetto dell'economia dei mezzi nei progressi dell'istruzione sociale, darei la preferenza, nei gradi inferiori della società, all'istruzione femminile in confronto della maschile, avendo maggiore influenza sull'educazione del popolo le madri, che non i padri.

Dopo tutto ciò, sono affatto del parere d'un mio amico, il quale allorquando associò una donna da lui amata e stimata alle sue sorti, le tenne presso a poco il seguente discorso: «Mia buona amica, quind'innanzi noi avremo tutto comune, i piaceri come i dolori, accrescendo i primi ed ajutandoci a sopportare i secondi condividendoli: ma però nelle nostre incombenze c'è d'uopo introdurre la divisione del lavoro. Noi due comporremo per ora, e fino a tanto che non sia giunto il tempo di assumere nei nostri consigli qualche principe del sangue, tutto il ministero della famiglia, di questo piccolo regno che ci è dato a governare. Tu avrai la direzione del ministero degli affari interni, io quella del ministero degli affari esterni. Dopo esserci bene intesi sulle massime generali di governo, ed avere bene studiato il capitolo delle finanze, ognuno agirà nella propria sfera indipendentamente dall'altro; bene inteso che talora si facciano dei consigli di gabinetto, nei quali, quando vi sia il bisogno, i due ministri facciano il loro rapporto. Io, come il tuo suddito, in ciò che riguarda l'amministrazione interna, sono dispostissimo non che a lasciar fare ad obbedire; sempre però proponendomi di far largo uso della libertà di parola, per il buon andamento della cosa famigliare, e talora anche di presentarti i miei gravami. Tu avrai gli stessi diritti rispetto al mio ministero, sebbene questo che tratta soppratutto, come dicono, delle *vie e mezzi*, non possa cosi facilmente divenire oggetto di quotidiana discussione. Vogliamo governare con buone leggi, facendo che l'autorità, la libertà e l'uguaglianza sieno aumentate dall'affetto; ed essere al caso di cangiare d'accordo la particolarità della Costituzione spontaneamente da noi accettata, ogni qualvolta faccia bisogno. Non esercitermo mai l'uno verso dell'altro le arti della polizia, condannata sempre all'ingrato ufficio di supporre il male, costretta da continui sospetti a cercarlo. Andremo piuttosto in cerca ciascuno noi di quello che può piacere all'altro. Non saremo gelosi, poiché laddove c'è gelosia, può albergare la passione, ma non l'amore. Il giorno in cui l'una o l'altro divenissimo gelosi, di qualunque fosse la causa, sonerebbe l'agonia del nostro amore, che nato quietamente e nutrito e cresciuto dalla azione dell'uno sull'altro, dal reciproco desiderio di piacerci e di soddisfarci l'un l'altro, rendendosi stimabili, dovrebbe durare per la vita.»

Vi risparmio, commaruccia mia buona, il resto della perorazione matrimoniale del mio amico, il quale dopo avere parlato dei difetti da correggersi colla reciproca educazione, o di quelli cui è di necessità il tollerare, terminava col mostrare in qual modo gli affetti di sposi e di genitori doveano in loro completarsi e contemperarsi coll'amore di patria, non sacrificando mai questo a quelli, ma piuttosto con essi unificandolo. L'essenziale era di farvi conoscere come io tenga affatto consono alla natura della donna e rispondente ai suoi ufficj sociali, questo ministero della famiglia. Vedete, che le assegno una grande parte nell'umano consorzio, se pure non si deve dire la parte maggiore e più nobile.

L'individuo si potrà considerare come uno dei molti componenti lo Stato, come uno degli atomi che riuniti formano il corpo politico di tal nome. Ma la famiglia è già per sé stessa il vero e completo elemento della società. In essa i diritti sono uguali, perché condizionati dai naturali affetti e dall'esercizio dei doveri dalla stessa natura insegnati cogli istinti che pose nell'uomo; in essa v'è disparità di funzioni e concorso ad uno scopo comune; in essa continuità di durata, con sempre nuovi innesti, che sul vecchio tronco si fanno. La famiglia insomma è la società in piccolo; è la perpertua di lei rinnovatrice, quella da cui dipende il miglioramento dei costumi, o la corruzione di lei. Assegnando adunque la famiglia come principale campo d'azione per la donna italiana dei nostri giorni, mostro di fare tanta stima di lei, da affidarle in gran parte l'educazione civile e sociale, e quindi il rinnovamento della società nostra: bene inteso che l'educazione è mutua, e che, per rendere efficace l'azione della donna, anche l'uomo deve fare la parte sua.

Quando dico, che alla donna principalmente si compete il governo della famiglia, mi faccio strada a considerarla nella sua qualità di madre, ed a chiedere che per essere tale venga essa educata, e non per altro. Ma di ciò avrò a dire più specificamente.

Ma basta qui di far capire, che assegnando alla speciale cura della donna tutto ciò che sta entro alle domestiche pareti, non intendo punto di metterla in grado subalterno rispetto all'uomo, ma di accordarle quella che la fa essere veramente sua uguale, e per il vantaggio sociale che ne dee derivare, superiore. Intendo ch'essa possa aspirare a qualunque genere di studio e di coltura, subordinatamente a questo grande scopo sociale, alla qualità sua di madre e di educatrice formata dalla natura; che essa eserciti la sua influenza anche fuori della famiglia, colla potenza della bellezza, della bontà, della virtù, della gentilezza dei costumi, della coltura. Operando in questo modo, essa non proverà gli sfinimenti, le noje, il mal di nervi, il vuoto del cuore, che la rendono troppo spesso malcontenta della propria condizione sociale senza ch'essa medesima sappia indovinare il motivo: né troverà pari alla sua dignità la parte di Aspasia, di dama de' cuori; né si lascierà abusare quale strumento degli altrui piacieri, e corrotta, corrompere, infelice e causa d'infelicità e di vergogna, contribuendo alla degradazione del suo paese. Essa, coll'instinto che ha di colpir giusto, pensando di quali uomini la patria bisogni per rendersi prospera ed honorata nel mondo, saprà formarli coll'accordare la sua stima soltanto a chi la merita e col far vergognare di sé stessi que' mezzi uomini che non si spaventano della loro nullità, perché non sono più atti ad accorgersene da sé.

Pacifico Valussi

La Ricamatrice, X. 5 (1 Marzo 1857)

La donna italiana considerata in riguardo all'educazione civile e sociale

LETTERA TERZA

A Caterina Percoto,
Consideriamo, o Caterina, la giovanetta pudica e semplice nella sua bellezza, come fiore che inconscio sorge anche fra i dumeti ed attira colla soavità de' suoi profumi e fa ammirare la splendidezza de'suoi colori, e colla delicatezza delle sue forme comanda rispetto; e vedremo in lei un essere che esercita la sua azione sulla società, ispirando schiettezza e gentilezza di costumi ed educando la gioventù maschile a spogliarsi d'ogni rozzezza e brutalità, che spesso trovansi compagne alla forza, quando non sia dall'amore e dal pensiero domata. In quell'età, la donna, se non è deturpata dal vulgare civettismo, dalle schimmierie galanti onde certe Aspasie da dozzina, a cui mancando coll'età l'impero, si fanno maestre; in quell'età essa è una poesia vivente, e dite pure un angelo, giacchè, l'espressione comune e bene spesso la più vera e la più appropriata. Senza andare a scuole d'amore nei romanzi di mestiere, che friggono e rifriggono l'amore in mille diverse guise, quale è fra gli uomini lo spirito gentile, quale l'anima per poco monda di grossolanità, che non ricordi qualche simile apparizione della sua adolescenza, che rimase educatrice di tutta la sua vita? Dante che cantò colla meravigliosa potenza che gli è propria la sua Beatrice, non è il solo che ricevesse ispirazione al generoso sentire, all'alto intendere al bene fare da una di queste apparizioni. Quanto gelosi dobbiamo adunque essere noi, di conservare nelle giovanette, che

saranno spose e madri d'uomini cui vorremmo crescere degni alla patria, questa virtù di Dio eletta che le fa ispiratrici d'ogni bella e buona azione! Or quante volte l'educazione femminile tende ad avvicinare presso di noi la donna a questo tipo?

Considerate la giovane sposa, che si concede premio al costante affetto d'un uomo degno, il quale non si sfruttò in materiali e sozzi godimenti prima di offrirsi ad una che divenga la madre de suoi figli, consorte nelle gioje e negli affari, ajuto e scopo ad un tempo alle sue fatiche, compartecipe del proprio essere, anzi parte di lui la più cara. Vedete que' due cuori che battono insieme, quelle due menti che s'incontrano nel medesimo pensiero, e che cogli occhi più che non le parole se lo dicono, que' due volti che specchio continuo l'uno all'altro si atteggiano ad un unico armonico sorriso di pace, di gioja, di contentezza; e direte quanto importa per il bene loro, per quello di tutti, che uomini e donne si educhino tali da essere degni d'una vita simile. Sospetti crucciosi, gelosie impronte, turpi cupidigie, vili invidie non albergano in tale connubio: ma rese dall'affetto piacevoli fino le fatiche appena in altre condizioni, nella solitudine dell'anima sopportabili, si trova in quella pienezza di vita un compenso che franca da molte amarezze dell' umana esitanza. Ma ahimè quanti sono, che sappiano prepararsi una tale felicità, e che piuttosto non si adoperino a distruggerla fin da quando cominciano ad essere padroni di sé medesimi!

Considerate la madre, che versa sui frutti del suo seno, quella esuberanza d'affetto, che le venne dalle conscie dolcezze d'un lieto connubio; che vede fuse e rinnovate sul fresco viso del suo bimbo due sembianze umane unificate dall'amore; che indovina su quell'aspetto che vogliano dire e pianti e sorrisi, e grida e silenzj, i quali nulla per altri significano; che di se stessa cresce una nuova vita con cura delicata ad ogni altro incomprensibile; che spira, ministra d'una seconda creazione, la parola sulle labbra infantili e l'intelligenza nella mente che sa nulla; che educa istintivamente con quel misto di dolcezza e di severità, che a nessuna scuola si apprende e cui i sapienti possano ammirare ma non saprebbero mai insegnare: e commossa da questo spettacolo veramente divino, esclamerete: infelice colei, che essendo madre rinuncia volontariamente a questi ufficj. I più alti, i più nobili a cui la donna possa mai aspirare! Ma sono forse tutte le donne educate ad essere madri, od anzi molte non vennero esse dalla falsa educazione sviate da ques'ufficio, a cui lo stesso istinto le avrebbe condotte?

Taccio qui della sublimità di quelle tante donne, la cui vita è un continuo e spontaneo sacrificio; di queste espiatrici degli errori e delle colpe altrui, di queste più che virili eroine, le quali sono grande compenso alle debolezze di tante altre infelici e che si trovano in tutte le classi della società. La virtù dell'abnegazione che brilla in queste umili imitatrici di Cristo la considero piuttosto come parte di religione, che non quale frutto della sociale educazione. Considero la donna, che nella società gode la sua parte di bene, soffre la sua parte di male, la donna della famiglia: e confesso, che ci manca molto nella generalità per raggiungere l'ideale che mi sono fatto, quello a cui l'educazione della donna dovrebbe mirare, e da cui anzi sembra bene spesso allontanarsi a bella posta.

Nella colta e ricca società poco assai si cura di educare la donna che sarà sposa e che sarà madre, ed il più delle volte, quando essa diventa sposa e madre degna, lo diventa a malgrado dell'educazione ricevuta.

L'istruzione che si dà bene spesso alle donne è quella, mercè cui dovrebbero brillare in una società, dove ogni uomo cerca l'altrui non il proprio, dove si ha bisogno di quelle medesime apparenze esteriori, di quelle doti che fanno applaudire una virtuosa da scena. In questa società l'amore di due sposi, che abbiano passata la luna di miele, pecca di ridicolo. La moda porta ad una specie di comunismo, il quale quand'anche non vada fino alle ultime materiali conseguenze, pure agisce come un costante principio di corruzione. Una madre, che abbia in cima d'ogni suo pensiero l'educazione dei proprj figliouli e che si studii di farla da sé in tutto ciò che è possibile, invece che affidare i suoi nati in tutto a mani straniere, quasi la si accuserebbe di affettare costumi ed idee contrarie a quelle del mondo galante. Si vagheggia nell' istruzione delle giovanette il tipo d'una cantante, d'una ballerina. Suonare, cantare, danzare sono le belle cose a sapersi, sono, se vuolsi, ornamenti desiderabili della vita sociale. Ma accordare a questo la prima cura, rendere la sala del palazzo, la casa cittadina una succursale della scena, lodare più quelle che più somigliano ad una mima, ad una cantatrice, fare per gli uomini un occupazione permanente di ascoltare ed applaudire donne formate a questa scuola: oh! questo non è certo contribuire, né alla dignità della vita di famiglia, né alla civile e sociale educazione del paese!

Ora meno che mai è opportuno di dare alle donne italiane una educazione che tende a fare di esse tante civette, ed a mantenere e diffondere negli uomini quella mollezza di costumi, che non potrà certo giovare al nostro civile rinnovamento. Almeno per ora si dovrebbe educare la donna in guisa, ch'essa fosse indicatrice di costumi maschi alla gioventù, e che si formasse alle severe virtù della famiglia.

Né ad essere spose e madri si educano quando le si pongono in mano a persone che abbandonarono la famiglia; quando quelle che devono vivere nel mondo si consegnano, per una singolare contradizione, a chi fece professione di abborrirlo, e non può conoscere come si esercitino le virtù di spose e di madri. Ma ora m'accorrgo, che la mia lettera sorpassa la conveniente misura, per cui serbo di continuare in un'altra la mia conversazione.

P. Valussi

La Ricamatrice, X. 6 (16 marzo 1857)

La donna italiana considerata in riguardo all'educazione civile e sociale

LETTERA QUARTA

A Caterina Percoto.
O buona Caterina, io ho un progetto; e voi nella mia qualità di giornalista, o di ministro del progresso come direbbe il D. M.,[7] mi permetterete di fare dei progetti. Il mio progetto è alquanto ardito; poiché tende nientemeno che ad iniziarvi ad una cospirazione, che vorrei diffusa in tutta la società. Notate bene, ch'io sto per le cospirazioni aperte, alla faccia di tutto il mondo; giacché delle segrete non ne aspetto alcun bene. Avete, o donne, qualche reclamo da fare? Fatelo apertamente, e se mai voi temeste d'essere ciascuna la prima a mettere il campanello al gatto, farò io adesso questo sacrificio e leverò lo stendardo della ribellione.

Cospirate adunque d'accordo, o donne, di qualunque età voi siate, contro i matrimonii male assortiti, protestate tutte contro l'indegno sacrificio, che di voi troppo spesso si fa. Protestate contro i matrimonii, che non sono altro se non un affare d'interesse, in cui la dote è la cosa principale, ed un'anima umana, la sua felicità, la sua virtù viene considerata come un accessorio insignificante; contro quelli voluti soltanto da pregiudizii di casta, pregiudizii in cui non s'usa ormai confessare pubblicamente, ma che si mantengono tuttavia, ad onta del ridicolo che li copre; contro quelli che si fanno fra impari d'età, sacrificando pure donzelle in tutto il brio della loro giovinezza a vecchi libertini consumati nel vizio. Protestate contro tutti questi matrimonii e mettete al bando della colta società tutti coloro che li contraggono, che li promovono, che li tollerano. Finché tali matrimonii continueranno ad essere frequenti, finché quelle stesse madri che ne furono le vittime non si vergognano di desiderarli e promuoverli, finché si preparano di tal guisa una scusa agli adulterii futuri, non si può parlare, colla speranza d'essere ascoltati, della felicità nel matrimonio, dell'amore dei coniugi fra di loro e d'essi per i loro figliuoli. Le vittime si ribellano facilmente contro quelle leggi sociali, che tornano a tutto loro danno. Sacrificate ad una società corrotta, esse diventano alla loro volta strumento di corruzione, e terminano col portare in trionfo il vizio. Quante volte una ben nata donzella, appartenente a ricca famiglia, dovette invidiare la sorte d'una villanella, alla quale è dato di congiungersi al prescelto dal suo cuore? Qual meraviglia, se queste diventano poi scettiche dell'amore conjugale e se si abbandonano a passioni indegne di loro?

C'è bisogno, o donne, per la vostra felicità e per l'onor vostro, d'un ardito colpo di mano. Dovete affrontare il ridicolo che in una certa società incontrano due sposi che si amano per la vita; dovete persuadere tutti, che v'amate, che vi stimate, essendo ben conscie, che senza l'amore il matrimonio è una vera catena, una schiavitù, da cui non sarebbe possible emanciparsi senza il disonore.

Le antiche storie dei cicisbei e la prontezza con cui gli stranieri colgono ogni pretesto per caricare di vituperi il nostro paese, per non arrossire di dovere tanto ad esso e d'essergli ingrati, fecero alle done italiane in Europa una cattiva reputazione. Leggendo i loro scritti, cui molte di voi hanno il torto di preferire alle opere nostrane, perché più della lingua italiana conoscono le straniere; leggendo i loro scritti si ha frequente occasione di vedere quanta poca stima affettino di mostrare a voi, tenendovi quasi inette alla vita virtuosa della famiglia, a quella vita ch'è il fondamento dei costumi, per cui si formano le Nazioni grandi e potenti. Se dovessimo bene osservare in casa loro, forse non avrebbero molta cagione di vantarsi; ma con tutto questo voi dovete distruggere tutte le apparenze che vi sono contro. Non date mano di spose che a uomini degni, e quelli amate e siate altere del vostro amore, lasciate che lo sprezzo cada sopra quelle che fanno altrimenti. Allora voi sentirete di essere veramente emancipate, di poter molto sui vostri sposi, di essere loro uguali e di esercitare un'azione benefica su tutta la società. Questo gioverà per l'educazione civile e sociale del vostro paese, più che molte istituzioni.

È strana la parola, ma l'adopero per essere inteso. Di fianco all'*emancipazione* si trova da per tutto ai dì nostri l'altra di *reabilitazione*. A forza di cangiar nome alle cose si giunse a *reabilitare* tutto e tutti.

De Maistre[8] volle *reabilitare* il boja e l'inquisizione. Altri disse meraviglie del *Knout*[9] a di qualche altro equivalente. Si *reabilitarono* i Turchi; si proporrà di reabilitare ..... i ..... le ..... tutto quello che l'opinione credeva di aver irremissibilmente condannato. Voi, apertamente cospirando, dovete *reabilitare* il matrimonio. Non c'è quasi romanzo, o dramma, o commedia che venga di là, dove sanno farli, che non abbia nell'età nostra contribuito ad abbassare nella comune opinione il matrimonio. La singolarità della cosa meriterebbe di tentare il paradosso, e di far credere, che col matrimonio sta il vero amore, e ch'esso non è soltanto un'officina da fabbricare figliuoli. Se l'età nostra perviene a questo, può dire di aver fatto una rivoluzione sociale; e se le donne vogliono, possono farla, poiché data un'altra tempera alle penne, queste scriveranno a favore più che la prima non avesse scritto contro.

Tante spose nel matrimonio anche il meglio assortito trovano *un vuoto nel cuore*, e lasciano quindi in esso un'apertura alle passioni per distrazione, o perché furono male educate, o perché non sanno intendere che anche la vita delle ricche ha bisogno d'essere occupata talora in qualcosa altro, che in fare la *toilette* del mattino e della sera, nel fare e ricevere visite nojose, in cui appena la maldicenza fa qualche diversivo, nell'assistere agli spettacoli. Quale meraviglia, se una vita simile non basta a riempire l'esistenza d'una donna, in cui la facoltà senziente che prevale abbisognerebbe di molto da consumare? Ma una donna educata convenientemente deve sapersi trovare un'occupazione anche per le sue ore d'ozio. Delle occupazioni cui l'educazione femminile deve preparare alla donna ricca, dovrò intrattenermi più a lungo. La donna italiana specialmente ne vorrebbe meglio delle altre. Che se la sposa diventa madre, di quali occupazioni e distrazioni avrà essa bisogno? Assorta allora nell'altezza del suo ministero, essa potrà piuttosto temere di dover togliere qualcosa all'affetto del marito per dedicarlo ai figli, che non di trovare un vuoto nella sua esistenza. Se essa sentirà di non essere stata convenientemente educata qual madre, ripiglierà l'educazione di sé stessa, per farsi educatrice de' suoi figli; ed allora la sua esistenza sarà piena e felice.

LETTERA QUINTA

A Caterina Percoto.
La donna, o Caterina, ha nell'umana società per suo compito precipuo gli ufficii di madre. Quello che ne dice la Religione nella Genesi, dove mentre si assegna all'uomo il duro lavoro della terra ch'ei deve far

produrre col sudore della sua fronte, a lei si predicono i dolori del parto e la conseguente educazione della prole, ce lo fa conoscere l'istinto connaturale alla donna. Anche fra gli animali irragionevoli l'ultima ad abbandonare i suoi nati è sempre la madre. E quelle che poterono o dovettero rinunziare all'affetto di spose, più difficilmente rinunziarono all'ufficio di madri; e voi vedete tante donne che non hanno figliuoli, in tutte le condizioni sociali, farsi proprii quelli degli altri, e ciò non soltanto fra parenti, anche fra persone del tutto estranee. E se io rammento qui con tanta soddisfazione vi vidi fare sapientemente da madre agli orfani nipoti vostri, non lo tenete come un volgare complimento, ma come un omaggio alla donna, per quella facoltà di educatrice della tenera età ch'essa possiede.

Presso ai Romani, ch'erano sopratutto per maschie virtù ammirandi, a giusta ragione riferivansi a tutta lode delle madri i meriti dei figliuoli. Colei che mi diè vita, e che nella sua tenera ed instancabile operosità era veramente la donna della famiglia, soleva dire: che quando una madre nutrisse ed allevasse per bene la sua prole, avrebbe abbastanza bene adempiuto il suo ufficio di donna. E questa massima, cui la madre mia desumeva dalla dirittura del cuor suo, è indicatrice anch'essa, dicasi la parola, della missione a cui la donna è destinata nel mondo. Domanderassi ad un uomo, che sia prode nelle armi per difendere la patria da' suoi nemici, che provveda alla sorte del suo paese coi sapienti consigli, che serva co' suoi studii ad estendere le conquiste dello spirito umano sulla natura, che sia in qualche scienza dotto, in qualche arte esperto, che si getti sull'una o l'altra delle infinite vie aperte all'umana attività: dalla donna una cosa sola si potrà domandare, ch'essa sia madre. S'ella è tale, se oltre al nutrire ed allevare la prole, la educa e la mette sul buon sentiero, essa avrà fatto abbastanza.

La donna adunque è essenzialmente madre; è educatrice prima di tutto. La cura dei figli e di tutto ciò che li riguarda nella prima età le si appartiene. Di qui la necessità per lei di rafforzare il suo affettuoso istinto con appropriata educazione, di rendersi insomma colta a fungere l'ufficio di madre.

Ed è questa punto l'educazione, di cui siamo manchevoli in tutte le classi. È qui dove la dottrina e l'arte dell'uomo mancano; perché ei potrà ben parlare dei principii generali, ma non discendere alle pratiche applicazioni. Questi sono studii, che alle donne medesime si competono. E ne abbiamo di buoni: ma resta tattavia moltissimo da farsi. In ciò principalmente vorrei vedere esercitata la letteratura femminile. Non sono studii che possano essere fatti da quelle donne che mandano fin da lontano un odore da pergamena accademica; non dalle Aspasie o dalle letterate per mestiere; ma propriamente devono esserlo dalle madri colte, a cui l'affetto materno sia luce dell'ingegno. C'è un vastissimo campo per la letteratura femminile, che deve anzitutto avere il carattere educativo; e per questo in Italia, più che altrove, resta molto da farsi. Bisogna porgere a tutte le donne, tanto della classe ricca, come della media e della povera, i mezzi per essere, oltreché madri, educatrici. C'è adunque un'intera enciclopedia di studii d'applicazione da comporre e da adattare alle varie condizioni sociali, per l'educazione fisica, intellettuale a morale delle madri, per ajutarle ad essere degnamente tali. Penso, ripeto, che questo sia lavoro da intraprendersi dalle donne medesime; ma se anche la letteratura maschile volesse dar loro coraggio a mettersi su tale via, potrebbe pure iniziare una *enciclopedia delle madri*, in cui si procurasse di porgere ad esse tutti gli ajuti per istruire sé stesse ed educare i figliuoli. Si prenda ispirazione dalle donne medesime; si faccia una raccolta delle migliori opere di donne di questo genere di studii, tanto nostrali che straniere; si procuri di colmare le lacune che rimangono, e si traccino ad ogni modo le prime linee di un lavoro complessivo, che serva di guida alle donne stesse.

Supponendo, che questa *enciclopedia delle madri* debba col tempo venire composta dalle stesse donne più colte, io vengo implicitamente a stabilire, che quelle della stessa classe più ricca debbano essere istruite meglio che ora non siano, affinché riesca qualche ingegno privilegiato atto a tanto. Troppa incompleta istruzione è tanto quella che si dà nei conventi e nei collegi, dove le educatrici hanno fra tutti il grande difetto di non conoscere che cosa sia una madre ed una donna di famiglia; come anche l'impartita dalle solite governanti, le quali per lo più sono anche straniere, e sapranno anche molto, ma non educare donne italiane. Bisognerebbe adunque pensare almeno a fare delle maestre in un istituto centrale, che potrebbe essere stabilito a Firenze a motivo della lingua che vi si parla. Dirò in altra lettera qualcosa degli studii, che, per mio avviso, alle donne si convengono.

P. Valussi.

Giornale delle Famiglie 7 bis (1 aprile 1864)

Memorie di Convento

LETTERA PRIMA

... Amphora caepit
urceus exit...

S. Lorenzo 10 aprile 18...

Mia cara Amalia!
Sai tu che la tua lettera mi pone in un ben grave impiccio? Mi chiedi consiglio sul modo da tenere onde educare la piccola Marietta? Io, vedi mi credo assai poco idonea a tal'uopo. Primieramente io vivo così segregata dalla società, e su molte cose la penso in un modo così poco d'accordo con essa che la mia opinione potrebbe facilmente urtare le tue più radicate convinzioni. Poi libri di educazione, io non ne ho letti che pochissimi. Immagina che non mi è neanche mai capitato nelle mani il tanto famoso Emilio di Rousseau. Ora dunque che ti ho confessato la mia ignoranza adempirò così come posso al debito dell'amicizia e ti dirò con tutta schiettezza quello che mi è venuto in mente riflettendo alla narrazione che mi fai delle graziose bricconate della tua piccola e del piano che per correggerla pare abbiate adottato, tu e la sua educatrice. Ben s'intende che tu mi vuoi sincera e che non darai più peso di quello che meritano a queste mie povere ciarle. Le cose ch'io ti dirò le cavo dal cuore e dalla triste esperienza dell'educazione ch'io stessa ebbi la disgrazia di subire. Quando penso che tu allevata in famiglia e sotto gli occhi della tua buona mamma,

in fatto di educazione ricorri a me che fui posta a tribolare per sette lunghi anni in un convento di monache, mi pare quasi un' ironia. Ma noi siamo due povere donne, incapaci di lunghe ed acute discussioni, desiderose soltanto di ajutarci coi frutti della nostra esperienza; io dunque ti narrerò il mio male, tu fanne un confronto col tuo bene e forse uscirà un po' di lume che valga a guidarti nella dolce impresa che ti suggerisce l'affetto materno.

A dirtela schietta, dal racconto che mi fai della scena dell'ufficiuolo* e dalle risposte che ti diede la tua piccola, mi pare che ti sei già messa, tanto tu, come la maestra sur una via, che non è la vera. E la via stessa che tenevano le mie buone monache e anch'esse colle migliori intenzioni del mondo: e perché tu possa tirarne un giudizio che ti salvi dall'andare più oltre, invece di ragionamenti, io non farò che narrarti quei miei brutti sette anni. Un po' per sera, come se fossero tante visite, e non badare se tiro innanzi alla buona. Sono memorie antiche che getto sulla carta così come mi vengono, e spesso interrompendo le chiacchiere per attendere alla faccenda dei bachi vicino ai quali ti scrivo.

C.P.

LETTERA SECONDA

... Nata in una romita villetta del Friuli amavo con passione l'aria aperta e il verde dei campi. Dopo la morte del padre mio, correre a piedi delle colline o sulle sponde del torrente, perdermi nel folto delle biade, vivere di solitudine e di fantastici pensieri, era la sola consolazione ch'io trovassi a quel primo dolore che mi ricordo aver sentito assai profondo. Ma questa vita selvaggia e quasi abbandonata non poteva durare. Avevo dieci anni, la madre mia carica di figli e nelle ambascie di una economia in dissesto facilmente comprese la sua impossibilità di badarmi e risolse di mettermi in un convento, dov'era monaca una vecchia sorella di mio padre. Avevo veduto il convento; un'edifizio severo, lontano dal centro della citta, isolato con rade finestre sempre cieche e riparate da diversi ordini di ferri. A pian terreno alcune stanze che chiamavano parlatorj, le sole che fosse permesso di penetrare,

* La mia amica nella sua lettera ch'io adesso non so più rinvenire, in fra le altre, mi raccontava della disobbedienza della sua figliola e di alcune risposte assai piccanti ed ardite ch'ella aveva dato così a lei che alla sua bambinaja a proposito di un ufficiuolo che l'era stato regalato e che ad onta del loro divieto, ella aveva voluto portar seco alla chiesa.

nude di mobiglie, tranne poche sedie di legno assicurate al pavimento che stavano ai lati delle grate, finestre larghe e basse ingraticolate di ferri tanto fitti che solo nei quattro angoli permettevano l'uscita di un pomo o di una piccola ciambella, non già della mano. Ivi, quando mi conducevano a far visita alla zia, compariva un'altra figura di donna assai scarna, assai rigida, vestita di nero e circondata di tenebre. Alcuni eleganti rosarj, alcune picciole palle a diversi colori, e ciambelle di sapore e di forma affatto particolare, erano i suoi regali. Terribile l'idea di seppellirmi in quella prigione: pure un ragionamento nella mia testa avevo fatto. Mi stava dinanzi la vita; questo mare sconosciuto ch'io doveva varcare, questa via ignota ch'io doveva percorrere; darmi una mano e guidare i miei passi potevano coloro che l'avevano incominciata prima di me; e poiché la madre mia non era in caso di prestarmi questo pietoso soccorso, chi meglio l'avrebbe potuto di quelle sante donne che si erano consecrate al bene con tanto sacrificio di se stesse? Mi avrebbero insegnato ad essere buona, avrei imparato da loro a lavorare, ad occupar utilmente il mio tempo, a fare quelle tante cose ch'io vedeva tutto di compiersi dalla madre mia, e che avevo sete di sapere. Alcuni anni di studio e di sacrificio e poi sarei uscita una donna e avrei potuto consolare, assistere, ed ajutare mia madre. In embrione capivo così che cosa fosse educare. Non capivo che quest'opera buona si potesse fare come un mestiere che vien compensato dal guadagno materiale; non capivo e ancora non arrivo a capire come per quest'opera di amore ci si possa valere della forza. Venne il giorno prefisso. Per non accrescere il dolore della separazione avevo promesso a mia madre di non piangere e mantenni la mia parola. Giunta dinanzi al mesto edificio udii cigolare i raddoppiati catenacci e stridere nella toppa le chiavi pesanti, e si spalancò la porta che doveva inghiottirmi. Sulla soglia era la zia, la madre Abbadessa, la Portinaja e facevano ala in due lunghe file, le quarantaquattro fanciullette che dovevano essermi compagne. Chiusero, e le prime parole che mi s'intimarono e ch'io allora non compresi furono, che la mia volontà da quell'istante restava fuori del convento. In grazia del mio ingresso quello era giorno di vacanza per le dodici ragazzine della mia classe. Consumammo l'intera giornata a visitare il locale; mi conducevano negli orti, ne'cortili; mi mettevano a parte dei loro giochi. Percorrevamo i lunghi dormitorj delle monache, l'infermeria i refettorj, i porticati che circondano la corte quadrata nel cui mezzo la apre la Fontana co' suoi puliti margini di pietra ombreggiati da un bel pergolato. Ivi sedute mi discorrevano degli usi del convento e della nuova vita che mi aspettava. Mi parve di notare che su molte cose esse la intendevano assai diversamente di me, ma ero troppo preoccupata dal dolore per por mente con riflessione a quel loro vivace cicaleccio. Aspettammo a visitare i nostri dormitorj, le scuole e la stanza da lavoro delle monache, quando la campana del Capitolo aveva già chiamata ai loro diversi uffici e doveri tutta la popolazione della casa. Subito che si entrava in qualche stanza dove fossero adunate al disimpegno della loro *obbedienza* o monache o converse od educande tutti gli occhi erano rivolti sopra di me e le prime si facevano debito, dopo diverse interrogazioni, di farmi un po'di predichina ch'io accettavo con affetto e della quale con tutta l'anima mi proponevo di approfittare. Mi parve d'accorgermi che le mie compange sogghignassero: ma dove non mi rimase più dubbio fu nella scuola così detta delle *mezzane*. Avevamo tanto tirato in lungo questa nostra escursione, che quando mi condussero a quella scuola era già vicina l'ora della ricreazione. Appena entrata la maestra che voleva vedermi da vicino e che allora era occupata ci fe' segno colla mano che ci assidessimo sulle panche presso la porta che servivano per le lezioni della calligrafia e che ivi si aspettasse. Ebbi campo di osservare e mi ricordo tutti i particolari di quella scena come se jeri fosse avvenuta. Il sole, già vicino all'occaso, mandava un lieve riflesso di luce rossastra al sommo delle invetriate a ottagoni di due grandi finestroni gotici dinanzi ai quali siedevano aggruppate intorno a due rozzi tavolini una ventina di ragazzette dai quattordici ai quindici anni che recitavano l'ufficio della Vergine terminando il loro còmpito. Esse erano vestite presso a poco tutte alla medesima foggia e pettinate coi capelli divisi in treccie ed annodate al sommo del capo, mentre le piu tenere d'età, quelle della mia classe le tenevano libere giù per le spalle. Alcune curve sul tombolo cucivano affrettate come se avessero temuto di non giugnere in tempo, altre colla faccia più ilare già spuntato il lavoro riponevano le spille, il ditale, le forbici e guardavano alla maestra che con delicato magistero andava compiendo alcuni fiori artificiali. In questo mentre s'aprì la portiera in fondo alla scuola ed entrò un'educanda ch'era stata in parlatorio. Ell'era oltre modo giuliva e piegato in gran fretta il grembiulino delle feste lo ripose sur'uno dei tanti cassettoni disposti in riga lungo le pareti della stanza; poi senza parere e guardandosi in torno fe' scivolare un piccolo involtino che teneva nascosto di modo che andasse a cadere dietro la panca sulla quale eravamo assise noi altre, indi andò al suo posto e invece di salmeggiare sussurrava sotto voce qualche cosa di assai lieto alle compagne che le stavano dappresso e che tosto si comunicavano tra loro in maniera che rispondevano distratte ai versicoli, e la Monaca più volte guardò severa in quella direzione e con la mano impose silenzio. Quando fu terminato l'ufficio ella chiamò per nome quella educanda e le ordinò di consegnarle subito la chiave del suo cassettone; poi fatte alzare in piedi altre tre impose che per quella sera non dovessero uscire dalla scuola. Accettarono il castigo disinvolte e si ammicavano fra loro con certi risolini di dispregio che a me facevano male. Maestra e scolare così dunque procedevano di concerto? Era questo l'affetto, la confidenza, l'amicizia che secondo le mie idee dovevano reciprocamente professarsi? Non v'era più dubbio. Invece di darsi la mano e unire insieme tutte le loro forze per raggiungere la meta, esse si erano messe l'una contro l'altra e ahimè io avevo già veduto la guerra!

Quella prima dolorosa giornata finalmente si chiuse. Io avevo bravamente durato a superarmi fino al momento di andare a letto. La novità delle cose che vedevo, il correre d'uno in altro oggetto, l'allegria e il chiaccherio delle vispe campagne, mi avevano distratta e quasi tolto il senso del grande sacrifizio a cui mi era assoggettata. Ma quando entrai in camera, non ne fu più nulla di continuare ad illudermi. Una volta avevo visitato un'ospedale. Quelle due file di letticciuoli sì fittamente adossati, che a primo entrare sparivano affatto alla vista i vani tra l'uno e l'altro ma ne rendevano la sinistra impressione. Aggiungi la stanza assai angusta, bassa con una sola picciola finestra munita di doppia inferriata. Sedie non v'erano e non avrebbero potuto capire: a canto a ogni letto invece stava un piccolo bancuccio entro al quale si doveva chiudere la camicia e le calze, e riporvi sopra pulitamente piegati i vestiti, dal lato opposto alla finestra un inginocchiatojo col Crocefisso: in mezzo alla stanza un rozzo candelabrio di legno con candele di sevo. Presiedeva una monaca che ad alta voce recitava alcune preghiere che noi accompagnavamo assai distratte e intente invece alla facenda dello spogliarsi. Sovra ogni letto stava sciorinata la camicia da notte e una specie di cappa Bianca che appena cavato l'abito vi si doveva allacciare sotto il mento per precauzione, m'immagino, di modestia. Quando fummo tutte coricate la monaca venne letto per letto ad esaminare come avevamo riposte le nostre robe e a farci prendere la positura di convenzione per addormentarsi. Si doveva esser giuste nel mezzo del letticciuolo, colle coperte pulitamente aggiustate, colle mani sul petto come i morti sulla bara, ed ella dopo averci fatto recitare una massima eterna, col suo Crocifisso ci segnava sulle labbra una croce ch'era il suggello del silenzio per non potere più aprire fino al momento della sveglia. Mi ricordo che proposi fermamente di obbedire. Spento il lume e partita la monaca, le mie compagne invece rotto subito il divieto ciarlavano alla distesa e lungamente mi punzecchiarono e mi schernirono perché facessi coro con esse; ma io non aprii bocca. Neanche dormii. Avevo il cuore infranto e mi pareva un sollievo a poter finalmente piangere. Oh la mia bella cameretta di S. Lorenzo ch'io avevo cangiato con quest'orribile dormitorio in comune senz'aria, senza luce, senza quiete! e mi passava continuamente dinanzi agli occhi gravi di pianto la magnifica prospettiva delle mie tre ridenti finestre. Le colline, la campagna, i circostanti villaggi, l'aria fresca della notte imbalsamata di mille profumi, le armonie dell'ampia natura, lo stellato dei cieli e la luna viaggiatrice che mi discopriva al chiaror de' suoi raggi il tremito della lontana marina! Seppellita come in una scatola di pietra in quella stanza centrale del convento che aveva per barriere prima il fabbricato, poi le altissime mura che lo circondano, indi le case cittadine, non avrei più mai veduto né il sorgere del sole, né il suo tramonto, né l'aspetto delle alpi gigantesche. Ad una ad una mi tornavano nel cuore tutte le memorie de' miei anni infantili. Vedevo come in tante visioni i noti volti de' miei cari e i luoghi amati dove io soleva menar la mia vita, ed ora mi pareva di correre in un prato insieme co' fratelli, d'inseguire le farfalle, di cogliere i mille fiori della campestre primavera, ora erano i baci e le carezze della madre mia. Poi era il mattino bagnato di rugiade col suo lume incerto, co' venticelli forieri dell'aurora, col tremolar delle frondi e lo svegliarsi degli augelletti, ed io col padre mio usciva a goderne le primizie e a beverne l'infinita armonia … Oh in quella notte io piansi mio padre con lagrime mai più versate e compresi che cosa fosse l'averlo perduto per sempre!

Caterina Percoto

Giornale delle Famiglie, 10 bis (16 maggio 1864)

Memorie di Convento

LETTERA TERZA

Mia cara Amalia,
Ripensando all'ultima lettera che ti ho scritta, mi viene in mente ch'io mi sono lasciata andare a raccontarti le mie vecchie storie di convento, senza farti un po' di preliminare per cui c'è il pericolo che tu mi fraintenda, o che certo gergo che senza accorgermi io vo adoperando tu non lo intenda.

E primieramente, quand'io ti parlo delle mie tante lagrime fra quelle quattro mura e della piega forse non buona che ricevettero in quei sette anni il mio ingegno e il mio cuore, non voglio che tu creda questa un'accusa o una tarda vendetta contro quelle povere e sante monache che mi volevano tutto il loro bene e la cui memoria mi è ancora così cara. Non era loro colpa, se la vita che si avevano eletto non si adattava alla missione di educare e d'istruire. Vi si erano sobbarcate per necessità, non per vocazione. Ti voglio narrare le origini del convento e vedrai che se inconscie di quel che facevano, tribolavano le giovanette cadute nelle loro mani, anch'esse erano alla lor volta crudelmente tribolate.

Nei tempi antichi, io credo verso il mille e duecento, un ricco patrizio udinese si pensò di fondare quel monistero. Sul muro esterno nel Capitolo c'è dipinta la sua immagine con una iscrizione che dice il millesimo di quell'opera pia in lingua romanza o volgare, cioè nell'italiano antico che forse allora anche qui si parlava. Il Capitolo è la parte più vetusta del convento. Serve adesso alle congregazioni delle monache; ci sono tre altari dirimpetto alla porta principale evidentemente eretti in epoca posteriore, una cattedra per la Badessa; all'un dei lati un'altro altare alla Vergine. Due finestroni gotici a vetri colorati danno luce alla stanza assai vasta che è come una chiesa, cioè d'un solo piano e un tempo isolate, mentre i claustri a colonati che fanno atrio e si prolungano a fiancheggiare il fabbricato che circonda la corte quadrata che sta nel mezzo del convento, sono addossati più tardi a questa specie di tempio che adesso dicono Capitolo e che forse era la sala di ricevimento del signor Occello Occelli, il fondatore. La sua immagine è un affresco antichissimo e forse ci rivela il costume bizzarro de'gentiluomini di quell' epoca nella nostra città di Udine. Egli ha in mano uno scudo a cuore su cui sta figurato un'aquila, ma con una sola testa che allora pare non sapesse dei mostri che noi nati più tardi abbiamo potuto troppo bene conoscere. Ha scarpe d'una forma così singolare da non poter capire come dentro vi possa essere costretto un piede umano; calzoni attilati e una gonna breve a guisa dei guerrieri antichi. Porta barba e capelli prolissi che riescono da un berretto rosso raccolto in mezzo da una gemma in forma di croce; il manto affibbiato sul petto a fiorami di vario colore rassomiglia al piviale dei nostri preti. Quest'Occello ch'è lì dipinto quasi sopra la porta a sesto acuto e tutta ghirigori dell'edificio, aveva una moglie che chiamava Verza, e una sorella Taraborella, pie dame a quel che dicono entrambe e di cui le monache fanno ancora l'anniversario con messa ed ufficio funebre. Non avevan figliuoli e fecero un'istituzione di tutti i loro beni per chi avesse voluto dividere con essi la loro vita di orazione e di ritiro. Legarono anche alcune doti perpetue per sei o più giovinette nobili di famiglia povera che la città trasceglie e che crescono in quel santuario, finché venga loro un'occasione di matrimonio o altrimenti abbraccino quella vita religiosa. Forse è questo legato il primo passo che trasse in seguito le monache ad assumere il difficile compito di dare educazione. Ma questo signor Occello faceva le cose in fretta e senza troppa previdenza perché la cronaca dice, che dopo aver disposto così di tutto il suo, gli nacque un figliuolo. Il padre lo battezzò col nome di Tardivenisti. Bisogna dire che i parrochi d'in quella volta non fossero schifiltosi come quei nostri che non vogliono saperne di nomi non registrati nel Calendario, o come quei vescovi che se capita loro alla cresima un'Italia o un Vittorio, lo cambiano subito in Giacomo, Tonio, Bella, e simili quasichè l'umanità dovesse sempre riprodurre gli stessi individui. Il signor Tardivenisti, non solo fu battezzato e anche cresimato con quel suo nome strambo regalatogli dal padre, ma pare anzi che nullaostante riuscisse un fior di un galantuomo e sopratutto un assai buon cristiano. E'si rassegnò al fatto del padre con tanta mansuetudine, che abbracciò anch'egli vita religiosa. Non è detto se si facesse prete o frate, ma in Capitolo c'è la sua immagine in abito da pellegrino col bordone e la cappa santa sul petto, e guarda pensieroso alla parete di contro; dove il pittore quasi a mito dei futuri destini del luogo ha raffigurata sant'Orsola e il suo seguito in una miriade di giovani testine tutte bionde e cogli occhi celesti che fanno piramide, nelle quali, a guisa dell'eroe di Virgilio o di Bradamante nella grotta di Merlino, il povero diavolo avrà forse veduto trasformarsi la sua possibile prosapia. Ora anche di questo Tardivenisti le monache fanno l'anniversario con messa funebre esequie ed ufficio, ed è sulla tabella dei loro benefattori, che in tale occasione espongono in coro, ch'io ho imparato il suo nome. – Coll'andare degli anni le case del Signor Occello che furono il primo embrione del convento, subirono di molte modificazioni. Parte caddero in rovina, e dal lato di tramontana c'è un vasto tratto di terreno irregolare confinante quasi colle mura della città, dove lavorando si trovano pezzi di terrazzo istoriati a mosaico e simili e altre anticaglie. Diversi corpi di fabbrica furono evidentemente aggiunti in epoche posteriori. Quello che costituisce la chiesa e i cori interni pare che sia costruito da ultimo, quando il monastero era già ricco e potevano spendere ed abbellire coll'arte. Gli affreschi sono del Quaglia di Como,[10] riusciti assai bene, particolarmente un gigantesco demonio che sta nel soffitto ed ha fra gli artigli una branca del pomo fatale. Nella chiesa interna o coro delle monache ci sono dieci grandi quadri a olio di buona mano e stucchi costosi e le cattedre di noce ad intaglio assai eleganti. Anche l'esterna è adornata di stucchi; anzi ogni affresco n'è contornato. Quella chiesa è una delle più belle della città benché un po' troppo forse carica di ornamenti e i dipinti abbondino di tinta giallo d'ocra, ch'era il difetto del Quaglia.

Ahi! ... m'accorgo che la penna m'è corsa e che t'ho detto non so quanti spropositi per voler entrare in erudizione che non sono alla mia portata ... Eppure ho da farti un'altra tirata, perché finora non t'ho detto che della parte materiale del convento, la quale in fin dei conti non è che il corpo destinato a custodir l'anima, cioè la istituzione religiosa che vive nel grembo di quel vasto casamento. E anch'essa coll'andare degli anni, a similitudine delle mura esterne, cangiò di forma e subì parecchie modificazioni. A principio non erano che pie dame amiche della Signora Verza e della signora Tamborella che si erano là ritirate e facevano insieme vita spirituale. Un po' alla volta cominciarono a legarsi con qualche voto. Alle ragazze povere che tiravano su presso di loro come tante figlie vi si aggiunse qualche altra o nipote od amica che si metteva nelle lor mani più perché fosse custodita e cresciuta nell'innocenza, che per darle un'educazione. Io credo che i nostri vecchi si rompessero assai poco il capo in fatto di sistemi educativi particolarmente trattandosi di donne. A parte qualche rara eccezione, qui nel Friuli facevano presso a poco come fanno ancora i nostri contadini. L'educazione dei loro figli consiste nell'imitare i genitori o gli altri della famiglia e col tempo subentrare a tutte le loro mansioni. Qui invece dei genitori e degli altri parenti erano le monache, e le giovanette che si davano loro in custodia avranno un po' alla volta imparato quel ch'esse sapevano e facevano. Vedi bene, che per tornar poi nel mondo e maritarsi e crescer su figlioli, non doveva essere il metodo migliore tanto più che queste monache dirette da frati di cui la città in quella volta abbandonava, pare che molto si occupassero di speculazioni religiose, o che almeno abbracciassero così a un di grosso le opinioni dei loro confessori e vi unissero agli stessi scopi le loro molte preghiere, perchè due volte ebbero un monitorio da Roma, che le richiamava alla purità della fede antica, come se si fossero lasciate strascinare a qualche credenza non abbastanza ortodossa. E anche col tempo e colle ricchezze cresciute, bisogna che avessero deviato dalla pietà delle loro sante fondatrici e che là dentro fosse accaduta qualche grosso disordine, perché Eugenio IV[11] credette bene di riformarle e di assoggettarle alla rigorosa regola delle clarisse. Dovevano pronunziare cinque voti, castità, povertà, vita comune, clausura perpetua ed obbedienza perfetta alla Badessa che si avrebbero eletto. C'e una lapide che dice il nome di quelle che abbracciarono cotesta riforma il che accenna che ci furono delle altre che non vollero assoggettarsi e che forse uscirono, o continuarono la vita a lor modo nelle loro casuccie perché c'è una parte del convento adesso disabitata ch'è come un aggregato di piccole cassette, dove forse prima del voto di vita comune vivevano separate le una dalle altre. E le cantine e i granai e i parlatorj e il grande refettorio sono fabbriche tutte posteriori a cotesta riforma di Eugenio IV. Il monastero ad onta del voto di povertà continuò ad essere ricchissimo per molti beni che possedeva qui e colà nel Friuli e particolarmente nel villaggio di Rizzolo che colle sue pertinenze era quasi tutto di loro proprietà. Vino, grani, ed altre derrate, tutto veniva al convento, e la Badessa col mezzo del suo amministratore vendeva, comperava, riscuoteva, e faceva in somma la padrona di casa. C'è un locale diviso in due stanze una interna e l'altra esterna comunicanti fra loro col mezzo d'una finestra munita di doppia inferriata, di dove si tengono i registri e la cassa e dove vengono a parlatorio pei loro affari la Badessa e l'amministratore. Quel locale di cui ella sola ha la chiave lo chiamano la casetta. La altre monache, senza espressa licenza di lei non possono tenere un soldo e non hanno di proprio neanche la tunica, così richiedono l'osservanza del voto di povertà. Anche dopo la riforma continuarono a tenere presso di loro delle giovanette, ma come già ti dicevo, piuttosto che in educazione, erano lì messe in serbo perfin che veniva un'occasione di matrimonio o altrimenti si fossero fatte monache.

La clausura rigorosa ivi osservata, teneva tutta quella popolazione femminile tanto divisa dal mondo, che là dentro assai poco si sapeva de'suoi progressi e di generazione in generazione continuarono sul piede antico finché capitò la rivoluzione francese e la guerra e la caduta della repubblica e Napoleone. Conventi di frati e di monache in Friuli s'erano moltiplicati fuor di misura, e il vincitore senza troppe discussioni, diede colla sua spada un taglio ardito e li soppresse alla bella prima. Fosse prepotenza di conquista, o frutto dei tempi e delle nuove idee; o fors'anco rimedio necessario a tornare in equilibrio il corpo sociale, certo la monache non sentirono che l'offesa recata alla loro libertà. Quelle di santa Chiara appartenenti alle prime famiglie del paese ricorsero ad ogni possibile protezione pur di parare il colpo e conservarsi nel loro nido. Riuscirono a restare, non come istituto di religione, ma sotto l'aspetto di un collegio per l'educazione delle fanciulle del quale nella provincia c'era assoluta mancanza. Dovettero peraltro cedere i loro beni e contentarsi d'una meschina pensione vitalizia passata dal governo, dovettero spogliare i loro abiti monacali adottando invece dei veli maestosi e delle sacre bende una semplice cuffietta bianca, e invece della tunica e dello scapolare, una povera vesta di lana nera. Sulla facciata esterna del monastero furono tolte la immagine di santa Chiara e le simboliche mani stigmatizzate di San Francesco ch'erano la loro insegna, e sostituito a lettere cubitali come quel sito non era altro che una casa di educazione femminile. Tutte le monache che non potevano prestare un ramo d'insegnamento vennero obbligate a sloggiare, e parecchie della malcontente colsero quest'occasione per tornare alla vita del mondo. La maggior parte peraltro rimase, anche di quelle fra esse la cui vocazione era stata forzata. Dopo tanti anni di reclusione, senza più la speranza di una famiglia propria, prive d'appoggio, abbandonare le compagne; abbandonare la loro cella, la loro chiesa, i luoghi testimoni forse delle loro lagrime, ma diventati cari per una lunga consuetudine, e tutto questo per avventurarsi a una vita sconosciuta in mezzo a gente che le avrebbe facilmente derise come una turba di nottole snidate di pieno giorno e fuggenti in una luce che le uccide, era per esse troppo grande patimento. (Continua)

Giornale delle Famiglie, 11 bis (1 giugno 1864)

Memorie di Convento

LETTERA TERZA

(Continuazione, vedi numero precedente)

Fu dunque allora che quelle povere monacelle che fino al quel punto non avevano si può dire vissuto che per la preghiera e per la meditazione assunsero di trasformarsi in tante maestre. Diedero in nota al municipio il loro nome che per insegnare a cucire di bianco, chi pel ricamo, chi per i rammendi delle calzette, una pei fiori artificiali, una per la lettura; la mia zia che voleva ad ogni patto restare e che non aveva saputo trovar fuori nessuna cognizione che la costituisse nel novero di coteste maestre improvvisate, così vecchia di già oltre cinquant'anni si mise ad imparare la calligrafia non essendoci là nessuna che sapesse scrivere il suo nome con un carattere regolare; e tanto fu il tuo desiderio di riuscire, che con poche lezioni venne approvata per cotesto ramo d'insegnamento. Così la scuola di musica venne affidata ad una monaca oriunda di Padova che nella sua gioventù, dicono toccasse assai bene il piano-forte, ma che dopo cinto il velo, non ne aveva più voluto sapere, siccome di vanità affatto secolaresca. Per le lingue e per il disegno furono obbligate a ricorrere a maestri di fuori che venivano a dar lezioni due o tre volte per settimana ed erano pagati dalle famiglie. In tutte quelle quaranta e più monache non c'era chi sapesse una sola sillaba né di francese, né di tedesco, o fosse in caso di segnare colla matita un arabesco di sua fantasia. È inutile di dire che la grammatica in quell'epoca là dentro non la conoscevano neanche di nome. Chi meglio disimpegnava il suo ufficio era la maestra di geografia, una vecchia veneranda cieca già da molti anni ... Non ridere! ma quand'io la conobbi, ell'era un vero portento. Suor Maria Elisabetta non era stata educata là dentro. Rimasta orfana con una sorellina di tre anni, ottenne d'esservi accolta insieme con essa. Amava con passione la lettura e fu più d'un poco tribolata dalle altre consorelle che vedevano in ciò una tendenza piuttosto mondana. Perduta la vista, fu confinata nella sua cella e Dio lo sa di che amaro calice venne ivi per lungo tempo abbeverata! Quando il monastero dovette trasformarsi in collegio, s'accorsero ch'ell'era un tesoro e correvano a consultarla. A' miei tempi teneva nella sua stanza una piccola biblioteca per suo solo uso, cioè erano libri che si faceva leggere da qualche ragazza o da qualche monaca che fosse stata di sua confidenza, e adesso, dopo molti contrasti finalmente glieli permettevano, come pure le avevano restituito un globo, una sfera armillare e alcune carte geografiche che vedevi appese alle pareti. Ci mandavano da lei ad ore determinate non già tutte, ma quelle poche soltanto che la Badessa trasceglieva. Il suo conversare era così caro, sapeva inspirare un tal desiderio d'istruirsi che questo solo valeva un'intera educazione. Povera Maria Elisabetta! mi par sempre di vederla ritta e dignitosa della persona, dolcemente sorridente, serena e rassegnata la faccia, tollerante come un angelo delle altrui ignoranze. C'insegnava così discorrendo i principj della geografia ed aveva così vivace e precisa la memoria, che quando noi non sapevamo rinvenire sulla carta un punto qualunque disegnato della lettera o dalla conversazione, ella se la faceva mettere dinanzi e chiestoci di posarle il dito sul tale o tal'altro grado di latitudine misurava colla mente le distanze e cieca toccava il punto che noi cogli occhi avevamo indarno e a lungo cercato.

A completare l'idea che tu adesso devi esserti fatta di questa nostra educazione di convento, basterà ch'io ti metta qui l'orario al quale eravamo assoggettate. Nei mesi estivi venivano a farci alzare a' primi crepuscoli, nell'inverno quando ancora era tutto bujo e al suono della sveglia accendevano i lumi. La monaca destinata a quest'ufficio intonava una preghiera durante la quale dovevamo tutte balzar dalle coltrici. Ella infilava la fascetta alla prima che obbediva, questa alla seconda e così di seguito in modo che nel mezzo della stanza si formava una specie di catena. Vestite, si scendeva in iscuola a pettinarci l'una l'altra. Fuori della scuola stava un secchiello e un asciugamani che doveva servire per tutte quattro le classi. Puoi credere ch'era un lavarsi assai spiccio, ma a cotesto non badavano le monache, le quali nella lor vita mortificata e tutta di annegazione in fatto di pulizia avevano certe idee assai singolari. Pensa che ce n'era una fra esse che ci additavano siccome modello di santità la quale da tre anni non aveva toccato acqua e poi c'era la sera del sabato in cui ci veniva concesso la lavanda del collo, solennità che mi riserbo a descriverti un'altra volta. Al tocco delle otto in coro alla messa, poi si attraversava la corte quadrata e si veniva in cucina, dove la campagna ci chiamava per la colazione. Finita questa, suonava scuola e consisteva in ogni sorte di lavori dall'umile calzetta ai più difficili trapunti che si eseguivano recitando l'ufficio della Vergine od ascoltando la lettura di qualche libro per lo più divoto. Ognuna aveva in consegna il proprio col suo indice fra le pagine e la maestra ordinava chi dovesse divertire le compagne col cotesto esercizio ch'era quasi tutta la nostra istruzione. Talvolta si recitavano così lavorando altre orazioni come novene, rosarj, ecc., ecc. Spesso ci veniva comandato qualche giaculatoria a nostra scelta ti assicuro che ne uscivano di veramente graziose. Tra le altre mi ricordo della seguente ch'era solita recitare una mia cara compagna molto allegra tutte le volte che veniva il suo turno – Senza badare, se fosse il meriggio, o altra ora qualunque, pacificamente seduta al suo telajo ella intonava con voce chiara e spiccata:

Vado in letto
Coll'angelo perfetto
Colla Vergine Maria
Amen e così sia.

A mezzogiorno in punto a pranzo e prima una visita al santissimo sacramento. In tavola ben s'intende che non era permesso parlare. Durante le prime mense una monaca dal pulpito che stava in mezzo al Refettorio leggeva una predica o la vita di qualche santo; alle seconde, mentr'ella si cibava a freddo in compagnia di quelle ch'erano state di *obbedienza*, cioè le infermiere, le ascoltatrici o chi veniva dal Parlatorio, noi dovevamo star ferme al nostro posto ed occuparci con un libro o colla calzetta.

Quando parlo di libri devi intender sempre le opere che stavano registrate nel repertorio del convento, alle quali era assai difficile che venisse aggiunto niente di nuovo.* Dopo tavola di nuovo visita

---

* Se la memoria non mi fa falla erano presso a poco codeste: *Il Salterio. Il Fior di virtù. Il libro delle vergini. La vita di Giosafat. Un Robinson ridotto in dialoghi. Alcuni volumi di Rollin. Un compendio della geografia di Gutrie. L'imitazione. Rodriguez. Cattaneo. Valsecchi. Rosignoli. Il Leggendario dei santi. Il Giovanetto Giuseppe. Altre opere di Padre Callino. La Filotea. La Manna dell'anima. Il Libro delle sante piaghe. Lo specchio della vera penitenza del Passavanti* e credo anche *Gli esempi del Prato Fiorito.*

in coro, indi un'ora di ricreazione, al termine della quale era pronta la campana a chiamarci un'altra volta in coro alla recita di non mi ricordo più quali preghiere. Nelle ore pomeridiane la scuola come al mattino, senz'altra interruzione che la sospirata cesta della merenda. Al tramonto la seconda ora di ricreazione in coro alla recita del Rosario, finito il quale portavano i lumi sul banco e si doveva fare un'ora e tre quarti di meditazione su un libro d'ascetica; mi pare certo che fosse la manna dell'anima. Si passava indi in scuola e invece delle solite maestre che restavano in coro a salmeggiare, veniva adesso una sorella conversa o qualche monaca malaticcia; e quel tempo che si stava lì aspettando la cena dicevano Fila, ed era l'ora che permettevano di studiare le loro lezioni a quelle di noi che imparavano le lingue e la musica. Immaginati che bel momento! Lì tutte insieme, con due sole candele di sevo per camerata in mezzo agli strilli della spinetta e alle chiacchiere di chi agucchiava o cuciva. A cena silenzio e lettura come a desinare. Poscia nelle nostre camerate è permesso il gioco fino alla venuta delle monache che ci conduceva in chiesa alla recita di altre orazioni, dopodiché era finita e si andava a coricarsi come ti ho raccontato nell'altra mia. La estate c'era una variante. Quell'ora detta di Fila veniva soppressa e invece lo studio delle lingue e della musica ce lo conducevano subito dopo la ricreazione del giorno, cioè dalle due alle tre, ora che chiamavano *il silenzio* e nella quale le monache potevano ritirasi al riposo nelle loro celle. I maestri di lingua francese e tedesca, e quello del disegno venivano a giorni determinati due o tre volte per settimana, e le ragazze che si facevano inscrivere per tali studj, non essendo obligattorj, ma soltanto tollerati, lasciavano la scuola e sotto la sorveglianza d'una monaca scendevano alla lezione nella camera così detta da ricevere. Mi dimenticavo di dirti ch'una volta al mese avevamo scuola di *belle lettere,* cioè dopo la colazione invece di sederci come di consueto a cucire, ci mettevamo in piedi ognuna dinanzi al nostro cassettone e lì si abborracciava una lettera, e con essa in mano attraversato la corte, una per volta si discendeva alla così detta cassetta della Badessa per la correzioni, e se la lettura era proprio da recapitarsi, si trovava in iscuola a metterla in bello, e spesso avveniva che i nostri sentimenti restavano in cassetta e in cambio si mandavano quelli della suddetta molta reverenda madre Abbadessa. Così pure le lezioni di calligrafia ci venivano date nel dopo pranzo una volta per settimana. Questo metodo quasi eguale per ogni classe, si manteneva invariato eccetto all'epoca degli esercizi spirituali; quindici giorni a carnovale, nei quali con grande nostro soddisfacimento non ci si occupava d'altro che del teatrino, di mascherati e di balli, cose tutto che intendo descriverti in seguito, e altri quindici o venti giorni prima degli esami autunnali, e allora bisognava mandare a memoria quella che chiamavano la grammatica, l'aritmetica, la storia e la geografia studi che nel resto venivano subito dopo posti per sempre nel dimenticatoio. Ma anche di questa solennità degli esami ti discorrerò un'altra volta, sicché, mia cara mettiti il cuore in pace che le mie lettere minacciano di moltiplicarsi all'infinito. Intanto per oggi addio,

C.P.

Giornale delle Famiglie, 12 bis (16 giugno 1864)

Memorie di Convento

LETTERA QUARTA

Mia cara Amalia!
Questa non è nessuna delle tre lettere che ti ho promesso, verranno più tardi: oggi invece m'ingegnerò di darti un'idea di alcune figure che devono entrare nel mio quadro e particolarmente di una ch'essendomi legata per parentela doveva avere od ebbe forse molta influenza così sul mio cuore, come sull'anima mia. Parlo della vecchia sorella di mio padre, ch'essendo monaca in quel convento fu il principale e forse l'unico motivo che indusse i miei a sceglier quel luogo di educazione piuttosto che un'altro qualunque. Sappi peraltro che da tempo immemorabile quasi tutte le giovanette nobili di cotesta provincia venivano educate là entro e che anche a miei giorni quell'istituto godeva di grande riputazione. Forse perché le monache appartenevano alle prime famiglie patrizie del Friuli e per entrarvi educanda anche allora ci voleva patente di nobiltà e solo soffrivano a malincuore qualche rara eccezione, causa i tempi mutati che le vecchie continuamente deploravano. Io che dal lato della madre puzzavo per almeno un quarto si sangue plebeo (mia madre era figlia di un fattore e di una gentildonna che lo aveva sposato in secondi voti) fui accettata con molta difficoltà e in seguito ben mi avvidi che non era possibile che mi perdonassero cotesta macchia originale. Nei giorni della loro benevolenza mi chiamavano la nipote di Suor Maria Gertrude e pareva che con questo titolo d'onore procurassero di generosamente coprire il disgusto invincibile loro prodotto dall'altra disgraziata metà

del mio sangue. Suor Maria Gertrude la mia zia, una delle anziane del convento, era molto rispettata. Donna di poche parole, di costumi semplici, di una pietà esemplare e in mezzo a quei tanti pregiudizj di un raro buon senso ell'era veramente rispettabile. Una delle poche che avessero abbracciata la vita monastica veramente per impulso del cuore si serbava fedele alla sua vocazione e scrupolosamente attaccata ai cinque terribili voti che aveva proferiti nella sua giovinezza. Ostia volontaria nel santuario del signore, il mondo più per lei non esisteva e trapassava solitaria in mezzo alle compagne meditando gli anni futuri e la santità del suo grande sacrificio. Io veggio ancora la sua alta e nobile figura, le forme gracili, gli atti severi; quella faccia pallida e impressa di soave mestizia, e la sua memoria mi commuove ancora a un senso di rispettosa tenerezza. Mi avevano raccontato de'suoi anni giovanili e della sua improvvisa risoluzione di cingere il velo. Nel nostro villaggio la generazione che l'aveva conosciuta era già scomparsa ma come una tradizione rimaneva ancora la sua istoria e io l'avevo più volte udita ripetere nelle case dei poveri contadini, particolarmente all'inverno nelle *file,* dove in grazia della vita di campagna e quasi selvaggia de' miei primi anni io spesso penetravo. Questa vergine delicata dalla pelle tanto candida che i raggi del sole non valevano ad offuscare, benché in abito succinto ed armata di moschetto seguisse instancabile i suoi cinque fratelli, il padre e gli zii alla caccia fin sulle più alte montagne del Carso, questa vergine singolare che tutti conoscevano in paese per averla veduta più d'una volta prender parte alle danze campestri nelle sagre dei circostanti villaggi e che poi tutto ad un tratto era scomparsa a quindici anni per ritirarsi nel fondo di un austero convento; questa vergine misteriosa era rimasta impressa nella loro memoria e narravano intorno a lei mille bizzarri aneddoti ai quali l'immaginazione avrà senza dubbio fornito buona parte del fantastico di chè andavano infiorati …

L'amica a cui erano dirette queste quattro lettere, in quel torno di tempo, o poco dopo venne a stabilirsi in campagna in un villaggio a poca distanza dal mio. Ci vedevamo quasi ogni giorno. Alcune signore nostre vicine convenivano in quella casa; quasi tutte erano state educate da monache: una fra esse nata a Recanati aveva passato dieci anni a Roma in uno di quei monasteri, e siccome l'affar che allora più c'interessava era l'educazione delle figliuole, così nel dopo pranzo, mentre queste giocavano insieme nel giardino o facevano coll'aja una passeggiata, si discuteva insieme il prò e il contro dei metodi o dei principi con cui eravamo state allevate. Tutte le lagrime di quell'età puerile, tutte le gioje e le speranze venivano ivi raccontate. Inutile dunque continuare le mie lettere; ma venutemi ora in mano e pensando che per dire alla mia amica dell'educazione ch'io avevo subita, tracciavo la storia di quei miei sette anni di vita e narravo i segreti di uno di questi luoghi di chiusura che l'epoca nostra ha preso ora in considerazione e minaccia distruggere, mi viene in mente che potrebbe essere opportuno continuare il racconto. Non più dunque in forma di lettere né allo scopo di rilevare i difetti o i vantaggi di una maniera di educare che la società sembra aver di già abbandonato, io ripiglio adesso la penna. Nel ritornare col pensiero al mio passato, più d'una volta mi pare ch'io ho vissuto, non una sola, ma varie vite. Cominciare, andar innanzi e poi per un gran dolore finire. Consumato questo dal tempo, rinascere, accogliere altri pensieri, rivivere ad altre abitudini, progredire fino ad un altro gran dolore. Questa è stata più volte la mia vicenda. La morte del padre mio e l'abbandono del mio villaggio per andare in città ad esser messa in convento è una di queste vite. La seconda affatto diversa sono i sette anni chiusi fra quelle mura. Mi si rivelò allora un mondo affatto nuovo, patii immensamente; un po' alla volta mi assuefeci, trovai bella la mia prigione e finii coll'amarla di modo che l'uscita fu quasi una morte. Monache e compagne, misteri di quel recinto inesplorato, gioje e pensieri di una solitudine inaccessibile quante volte nella fantastica vostra forma e con tutti i vostri bizzarri colori, non mi passaste voi per la memoria? Li metto qui sulla carta e saranno come una biografia di quell'epoca cara insieme e dolorosa al mio cuore!

C.P.

La Stampa,[12] XXXIX. 56 (25 febbraio 1905)

IDEE, PERSONE, E COSE

La Serva

La parola è brutta: ma non ve ne è un'altra. E quando dico serva, io voglio comprendere, in una sola brutta parola, tutta la bassa e alta domesticità femminile: cioè l'austera e imponente istitutrice straniera carica di diplomi, che conosce e insegna quattro lingue, che passa dalla figliuola di una principessa mediatizzata alla nipote di un duca inglese, e da questa alla ereditiera di un marchese spagnuolo: cioè la governante inglese, bionda, pallida undicesima figliuola nubile di un pastore protestante, che se ne va per il mondo, lavando con la spugna i bimbi e le bimbe altrui: cioè la bambinaia svizzera, con la sua persona massiccia, il suo *jersey* nero stirato sovra un busto quadrato e la sua canottiera col nastro: cioè la cameriera francese, bruttina, agile, furba, sveltissima: cioè la bambinaia, la cameriera, la domestica, la serva italiana con i caratteri spiccati delle regioni donde vengono. Tutte serve, costoro, anche quella che guadagna duecento lire il mese e il trattamento; tutte serve, dalla prima sino all'ultima, di ogni nazione, le *misses*, le *fraulein*, le *mademoiselles*, le *maids*, tutte quante, serve, non vi è altra parola. Che pranzino a tavola con i padroni, o all'*office*, o in cucina: che dormano con la giovinetta, nella più bella camera della casa, o in una stanza umida e oscura: che si vestano bene o si vestano male: che abbiano una cultura o siano delle ignoranti: che siano tenute come familiari o siano trattate con severità e con disdegno: tutte serve! Come chiamarle diversamente! Il loro carattere profondo e invincibile è il servilismo: niente arriverà a cancellare dalla loro vita questa espressione morale e istintiva: niente eleverà questa origine, questa inclinazione, questa tendenza, questa manifestazione: niente potrà trasformare questo carattere: e salisse sovra il trono una serva per un miracolo sociale, sotto il manto sovrano, sotto il diadema gemmato, ella trasalirà se ode suonare un campanello. Non vi sono eccezioni: non vi sono che apparenze diverse: il contenuto è sempre quello. Voi troverete una istitutrice che fa dei versi: voi v'incontrerete in una governante che ha l'aria di una dama di Corte: voi pescherete una bambinaia che sembra una suora di carità: voi cadrete sovra una cameriera che sembra un'attrice recitante Goldoni: voi sarete stupito, innanzi ad una serva che sa ricamare e suonar l'arpa. Illusioni, illusioni! L'anima servile è sempre in fondo a tutto questo, e immancabilmente ad un certo momento voi la scorgerete, viva, palpitante e fremente, in tutta la sua cruda realtà e in tutto il suo mistero pauroso.

Tu sorridi, amica lettrice? Tu non hai paura della tua serva? Hai torto. La più semplice prudenza sociale, la più infantile psicologia ti consiglia a diffidare della tua serva, qualunque essa sia, da dovunque venga, sia da venti anni in casa tua o vi sia da un giorno, abbia messo in pericolo, per devozione alla tua vita, la sua vita, o ti porti una tazza di camomilla fredda quando a te occorre calda. Non ti fidare. Non le credere! Non le credere mai! È una serva: e soffre atrocemente di essere tale: e la sua sofferenza accresciuta da lunghi anni, si è mutata in invidia livida, in collera sorda, in disprezzo celato, in odio, in odio possente e nascosto: ella è serva, e per questo odia tutti coloro che non sono servi: e ti odia, te, particolarmente, perché non sei serva, perché hai il diritto di comandare ed essa ha il dovere di obbedirti. Non credere alla mellifluità della sua voce; guarda bene i suoi occhi quando essa ti guarda e non sa di esser veduta: non credere al suo sorriso, quando entra nella tua camera, e ascolta bene, se ciò ti va, quando essa discorre in anticamera con qualcuno e suppone che tu sia lungi: non credere al suo rispetto e prova a far cadere questa corteccia sottile, udrai cose che ti faranno turare le orecchie per il ribrezzo. Ella ti odia. Se tu la compensi con generosità, dirà che sei una pazza, e non ti sarà grata: se tu le dai i tuoi vestiti, dirà che la vuoi umiliare e che essa non porta gli stracci altrui: se tu la soccorri nelle sue contingenze, dirà che lo fai per ipocrisia: se tu l'assisti nelle sue malattie, dirà che sei innamorata del medico. Ella patisce orrendamente di essere serva:

e ti odia. E la sua dimora presso te non è che un costante piano per portarti via tutto quello a cui tieni, un infernale piano di vendetta domestica, come se tu fossi la sua più grande nemica ed ella vivesse sotto il tuo tetto solo per distruggerti. Quello che tu ami, quello che tu prediligi, i tesori spirituali e i tesori terreni della tua esistenza sono in pericolo quotidiano e gravissimo per opera della tua serva. Hai tu una pelliccia rarissima? La tua cameriera la farà tarlare: e si stringerà nelle spalle , quando l'opera infame sarà nota. Hai tu una bimba che adori? La tua bambinaia, quando è alla passeggiata, la schiaffeggerà spaventandola poi perchè essa non te lo riferisca. Hai un figlio grande, giovine, bello, simpatico? La tua governante cercherà di sedurlo, o per isposarlo, o per farsi rapire da esso. Hai tu un marito che ami teneramente? La tua istitutrice, se è matura e non possa conquistarlo, farà di tutto per seminare la discordia fra voi due. Hai tu, diciamo la parola, un amico a cui hai dato il miglior tempo tuo e tutto il cuor tuo? La tua cameriera, se è appena passabile, tenterà di strappartelo: e, forse, vi riuscirà. Hai tu un nemico personale, acerrimo, con cui sei in lotta aperta? La tua domestica, allertata da lui, gli venderà la tua esistenza intima, lo introdurrà, magari, in casa tua e tu sarai perduta. Hai tu un segreto dolce e terribile da conservare? La tua anticamera lo conosce: lo conosce la tua vicina e domani lo saprà tutto il mondo. Sei tu la più infelice fra le donne, e vuoi che nessuno lo sappia? Il beccaio, il panattiere, il tabaccaio, sanno il numero delle tue lagrime. Hai tu commesso un errore? Colui che può puniterne su questa terra lo saprà dalla tua serva. Ti sei tu macchiata di un delitto? La tua serva ti denunzierà alla giustizia. Tutti potranno avere pietà di te, da Dio nei cieli sino all'estraneo che passa nella via: non colei che ha mangiato il tuo pane e che ha dormito sotto il tuo tetto. Pensi tu che io amplifichi, che io esageri? Rammenta: osserva: leggi!

È, dinnanzi ai giudici, in codesta grande e nobile Torino, una serva; la Bonetti. Si salverà dall'accusa di venificio, la Tosetti? Può anche darsi. Ma sia pure, condannata, altre cameriere avveleneranno le loro padrone, per prenderne il posto: e altre serve chiuderanno il padrone vecchio, malato, in casa, in camera, per averne il testamento o per averne la eredità. Lettrice, lettrice, leggi! La povera contessa di Montignoso, che poteva essere regina di Sassonia. Ed è una misera moglie perseguitata dal marito, che aveva cinque figli e non li vede da due anni, questa infelice donna che sconta così amaramente il suo peccato, quando altre donne ne fanno pompa e se ne gloriano e sono adorate, la contessa di Montiglioso non aveva, non ha che un sol bene, su questa terra: è la piccola Anna Monica, una bimba di venti mesi, con cui visse in solitudine, in una villa di Fiesole. Ma il Re di Sassonia le vuole togliere questo unico preziosissimo tesoro, e chi è mai la complice del Re di Sassonia? La bambinaia, Maud, la *bonne* di Anna Monica: Maud cioè la sua serva. La sventura della signora di Montignoso e la sua bontà: la tenerissima età della bimba e la sua grazia: l'adorazione di quella madre per quella bimba: nessuna di queste cose commoventi, che frangerebbero un cuore di pietra, hanno potuto vincere Maud, la *bonne*, la serva. Era lì per ispiare e ha spiato: era lì per tradire e ha tradito: era lì per denunziare e ha denunziato la sua padrona, quella con cui viveva, e ha firmato la denuncia, davanti ad un notaio! Era d'accordo per rubare la figlietta a sua madre, la bambinaia Maud: e ciò non le è riuscito: ma era di accordo per togliere alla sventuratissima contessa di Montignoso il supremo bene della sua vita. Di Maud, come tu vedi, o lettrice, e di tutte le sue gesta come spiona, come delatrice, come traditrice, sono zeppi i giornali; e essa si è dovuta mettere sotto la protezione di un Console a Firenze, perchè, forse, l'indignazione della gente l'avrebbe colpita. Che importa? Domani ella sarà obbliata; e il dì seguente ella andrà altrove, a fare del suo odio di serva uno strumento di altre vendette oscure o preclare. Maud sarà dimenticata: ma, domani, altre serve, di ogni classe, di ogni nazione, metteranno tutto il loro odio per perdere, per distruggere qualunque donna, sia la loro padrona, una Sovrana, una borghese, e per ricominciare, sempre, dappertutto, l'opera di perdizione e di distruzione. Serra i tuoi tesori, lettrice: chiudi la tua anima: metti ciò che ami in un tabernacolo alto, lontano, avvolto di ombra: sii tu, tu sola la custode di ogni tuo bene: e avrai salvato la vita della tua serva.

Napoli, febbraio 1905
Matilde Serao

(Siamo lieti di annunciare ai nostri lettori che abbiamo ottenuto la periodica collaborazione della prima scrittrice d'Italia, la quale la inizia con questa vivace requisitoria contro tutta una classe, che ha pure delle virtù e delle bontà qualche volta ignorate).

La Stampa, XXXIX. 60 (1 marzo 1905)

IDEE, PERSONE, E COSE

La Padrona

LETTERA APERTA A MATILDE SERAO

Cara Matilde,
Da parecchi anni non scrivevo più un articolo, e chissà per quanto tempo non ne scriverò ancora. Ci voleva il tuo nome sotto uno scritto che ha ferito profondamente il mio sentimento di giustizia, ed il caro ricordo della tua anima buona, tanto in disaccordo con la crudeltà delle tue parole, per farmi riprendere la penna abbandonata.

Ti sia questa una prova della mia antica e inalterata amicizia.

Tu hai parlato della serva. Io parlerò della padrona. Non per fare il processo alla padrona in genere, perché ne conosco tante piene di indulgenza, di evangelica bontà, come lo sei tu stessa. Ma osserverò i suoi rapporti con la serva ed il grado di influenza che l'una può avere sull'altra.

La padrona! La parola è ugualmente brutta, perché nel mondo evoluto, nella umanità umana, non vi dovrebbero essere né servi né padroni; ma soltanto creature uguali dinanzi alla natura e alla giustizia, creature che si amino fraternamente e si aiutino a vicenda.

Serva e padrona stringono un contratto. Una dà l'opera sua, obbedienza cieca, rispetto e sommissione, e tutte le sue ore del giorno e della notte, il che vuol dire la sua libertà. L' altra dà in compenso un salario, l'alloggio, il vitto, e la concessione di qualche ora libera, in un dato giorno della settimana. Non vi sono implorazioni né elemosine da nessuna parte. Sono pari.

Ma non è pari la loro sorte.

Vi sono padrone che esigono dalle serve fatiche superiori alle loro forze di donne! Intere giornate in piedi a stirare, salite dalla cantina ad un terzo o quarto piano con un carico di carbone ... Ve ne sono altre che misurano loro la porzione nel piatto, e tengono sotto chiave gli avanzi di credenza e le provviste.

Ve ne sono che trattano la serva con alterigia e credono molto nobile non rivolgerle mai la parola. Ve ne sono che le impongono un dato vestire dimesso, e guai se la serva, col frutto delle sue fatiche, si permette qualche fronzolo, povera concessione alle sue ambizioni giovanili.

Le serve debbono parlar sempre tra loro a bassa voce come in chiesa, o star zitte perché la padrona non abbia il fastidio di sentir chiacchierare.

Le contadine, avvezze ad allietar la fatica del lavoro colla gioia del canto, debbono rinunciarvi completamente appena entrano in servizio, perché la padrona non può permettere che la serva canti. E quando una serva è sola in una famiglia modesta, che noia deve subire nelle lunghe ore solitarie, sempre relegata in cucina, mentre sente nelle stanze padronali, ridere, discorrere, cantare, sonare.

Molte padrone quando vanno a teatro o in società esigono che la cameriera, o magari l'unica serva, stia alzata ad aspettarle fino a mezzanotte, fino al tocco, e più in là. La povera donna che è in piedi dal mattino, è costretta a lottare per lunghe ore col sonno, o, quando la natura la vince, a spasimare nell'incubo, a svegliarsi, in sussulto pel terrore di non avere udita la scampanellata della signora.

E la sensibilità raffinata della padrona non mette una punta di amarezza nel piacere delle sue serate, al pensiero di quella muta ignorata tortura che infligge ad una sua simile?

E se la serva, nella sua anima immortale, sente sorgere le ribellioni che il confronto dei due diversi destini può suscitare, non è ovvio, non è umano? e, sopratutto, non è un sentimento opposto all'idea di servilità di cui tu accusi tutta quella classe di diseredate?

Vi sono serve maldicenti, bugiarde, invidiose, ladre anche; ve ne sono di omicide, come quelle che tu citi. Ma non tutte le maldicenti, le bugiarde, le ladre, sono serve.

Signore, che ci chiamano amiche, sparlano alle nostre spalle quanto le umili serve. Lo faranno con maggior spirito, con arte più raffinata; ma la perfidia è la stessa.

E le omicide, le avvelenatrici non sono sempre serve; molti, troppo fatti recenti lo provano. Le stesse serve che tu citi per la vergognosa notorietà che risulta loro da scandalosi processi, furono incitate, spinte, trascinate alla nefanda impresa da mariti, da fratelli delle vittime, i quali non debbono avere un'anima servile.

Ma le serve come le padrone delinquenti – perchè vi sono anche le padrone delinquenti, – sono casi eccezionali.

Quella che tu stabilisci come regola e causa di ogni malvagità è la servilità nell'animo delle serve. Ma i torti di cui le accusi sono piuttosto una ribellione del loro animo alla servilità della loro situazione. « *È serva e soffre atrocemente di essere tale* » tu dici. E non ti pare che sia questa una manifestazione del sentimento dell'uguaglianza umana, che spasima e protesta contro le ineguaglianze della società? E ti pare che queste proteste disperate siano di anime servili?

Lascio indifesa l'accusa di subdole cospirazioni con le tarme per far tarmare le pellicce della padrona. Non conosco tanto a fondo l'anima oscura di quelle bestiole per poter misurare la loro parte di responsabilità sul delitto; ma mi sono grandemente sospette perché loro sole ne traggono vantaggio.

E quando la serva, giovane, forse bella, qualche volta pura – almeno per il primo fortunato – diventa l'amante del marito o del figlio della padrona, tutta la colpa è proprio di lei stessa, della sua anima servile? Quei poveri uomini sono stati colti per sorpresa? Violati? E quella ingenua padrona ignorava che la convivenza nella stessa casa, a tutte le ore, di una ragazza con uomo giovine – o anche mezzo vecchio – è una tentazione ed un pericolo? Quanta sventura! E quanti tradimenti contro quelle anime non servili!

La padrona può avere, come tu supponi un ..... amico «al quale ha dato il suo miglior tempo e tutto il suo cuore, e la cameriera tenta di strapparglierlo». Questo è un atto crudele, sleale, è un tradimento ..... Ma se ne fosse ardentemente, pazzamente innamorata, quella cameriera? (Anche le serve possono amare). E tu hai dimostrato nei tuoi romanzi così vissuti, così suggestivi che l'amore si impone a tutto, che non si lotta colle passioni. Per amore le amiche hanno tradito le amiche, le sorelle hanno tradito le sorelle, come in quello splendido *Cuore Infermo*.

Il tuo ingegno brillante e versatile, ha voluto far dello spirito alle spalle delle serve.

Ma tu sai che siamo tutte figlie d'Eva, tutte soggette agli stessi errori, tutte capaci delle stesse abnegazioni, tutte colpite dalle stesse fragilità, tutte meritevoli delle stesse indulgenze.

E tu che hai versate così sante lacrime su la via della Croce, *Nel paese di Gesù*, sono certa che trovi nel tuo cuore di fervente cristiana tesori di fraterno amore e di infinita pietà per alleviare alle nostre sorelle sfortunate il peso e l'umiliazione della loro sorte.

Torino, 27 febbraio 1905

La Marchesa Colombi

L'Illustrazione Italiana,[13] III. 23 (2 aprile 1876)

La Donna Libera

Idee di Neera

Io non sono per l'emancipazione della donna, ecco, lo dico addirittura. Nel mio umile parere trovo che la donna è sufficientemente emancipata; secondo me, lo schiavo è l'uomo.

Prendiamo un esempio nella classe media che è la più popolosa e la più influente.

Il fanciullo a dieci anni è obbligato di mettersi seriamente a studiare.

– Scegli, gli dice il babbo, fra il latino e le matematiche.

– Non mi piace né l'uno né l'altre; preferisco il cervo volante.

– Non è possibile; devi prendere una carriera, devi essere avvocato o ingegnere, dottore o professore.

Colla bambina non si è tanto esigenti. Ella studierà se ne avrà voglia – Alla fin fine – dice la mamma – non dobbiamo farne una letterata.

Pierino deve, ad ogni costo, essere un uomo di talento. Mariuccia non ha che l'obbligo di essere bella.

A vent'anni la fanciulla ricama presso sua madre e non le è vietato di mostrarsi qualche volta alla finestra. Abita rimpetto un bravo giovane, giudizioso e ricco – non si sa mai !...

– Ma sorveglia un po' nostro figlio, dice la moglie al marito; bada con chi pratica; non vorrei s'invischiasse in qualche amorosa pania – non è la sua età di pensare a tali cose (il giovinetto è quasi sempre d'opinione contraria); ma la prudente mammina che non farebbe nessuna difficoltà a maritare la fanciulla – anzi si spaventa al solo pensiero che quel ragazzo voglia prender moglie.

Fresca, elegante, cinta di fiori, la giovinetta va al ballo; l'accompagna con ansia amorosa la fida genitrice che non la trova mai bella abbastanza – un po' più indietro, questa rosa – più in giù la cintura – allaccia il quinto bottone del tuo guanto destro – dà qui, lo allaccerò io.

All'indomani della festa la giovinetta è deliziosamente immersa nelle memorie dell'ultimo *galoppe*, che, fosse ballato da sessanta coppie, impone l'obbligo a sessanta ballerini di pronunciare questa frase languida:

– Ah!... signora, come mi è sfuggita la notte ... avrei voluto che durasse una eternità! Danzando con lei sembra di volare in cielo ...

La madre è pensierosa e si affaccenda attorno ad una piccola valigia – il figlio parte – è soldato. Non va volentieri; al contrario, detesta la vita militare; egli ama la sua famiglia e la tranquillità domestica. Che serve? È soldato – deve andare. Dove? – non si sa. Il momento dell'addio è giunto – tutti piangono, egli no – ma soffre più che tutti. Abbraccia la sorella e avrebbe voglia di esclamare:

– Tu almeno sei libera!

I genitori sono morti. La fanciulla è sposa e madre; ha la sua famiglia, ha nuovi esseri che ama di sviscerato amore – è padrona della sua casa – comanda – il suo piccolo regno pende da' suoi cenni.

Il giovinotto ha trovato un impiego. Non è più tanto un giovinotto, a dir vero – sarebbe ora d'accasarsi. Egli lo desidera, oh! molto – ma come si fa? Guadagna mille e duecento lire.

Ha conosciuto una donna – riamato l'ama. Si parlano, si scrivono, sospirano – a quando le nozze?

Ma! ... – il principale non vuol crescere lo stipendio – gli impegni di una famiglia sono molti. – Se un terno al lotto! Il terno non viene.

La fanciulla, cui fuggon gli anni, sposa un altro. Egli resta solo, disilluso, infelice. Ha il conforto, è vero, di essere elettore, di essere giurato, di servire nella guardia mobile e godere di tutti i diritti civili.

Se è questa l'indipendenza che manca alle donne – grazie – la mia parte ve la regalo.

Uomini di buon volere (badate che il buon volere non lo nego), animati dalle migliori intenzioni (obbligatissima!), hanno pensato di schiudere alla donna la via degli impieghi. Secondo loro, dovremmo avere la donna telegrafista, la donna contabile, la donna *impiegato*, insomma. Parte io pure di quel sesso che è tanto felice di sentirsi chiamare debole, mi si permette di esprimere un'opinione? Sapete che io sono franca e molto veritiera.

Ebbene, la donna è nata per piacere agli uomini, per propagarne la specie, migliorarla, ingentilirla e far calze. Io non le riconosco altre missioni e mi pare ve ne sia abbastanza. Togliete una donna alla casa, e non avrete più nè casa nè donna.

Le condizioni del vivere sono diventate così difficili – dicono – che bisogna schiudere un avvenire alle povere donne rimaste zitelle per non aver potuto trovare un marito in grado di mantenerle.

Volete sentire una favola? Giuro che non è di Esopo e nemmeno di Lafontaine.

« Un eremita aveva una gallina, ma egli era tanto povero, tanto povero che non le poteva dar nulla da mangiare. – Mandala fuori per il mondo – disse un ricco filantropo – io le darò tutti i giorni il suo piattello di riso. – Ah! dà a me il piattello e lasciami la gallina – rispose il pover'omo. »

Avete capito? A me pare che avesse ragione.

Se la strada degli impieghi è ristretta, se rigurgita di aspiranti, perché aggiungervene altri? Che razza di economia sociale! Per tante donne che avrete messe a posto, vi resteranno altrettanti uomini disoccupati – dov'è il vantaggio?

Voi dite: La donna che guadagna troverà più facilmente marito. Ammetto; ma credete sia un vantaggio per lei? Che marito sarà quello che specula sul lavoro della moglie? Che risultati avremo da questo invertimento dell'ordine sociale, antico come la civiltà e basato sulla diversa natura dei due sessi?

La donna lontana dalla casa non sta bene. Le abbisogna il focolare colle sue gaje faville e coll'amico caffè che gorgoglia fra le bragie – le abbisogna il noto panchino – la vecchia e comoda seggioletta a bracciuoli – il canarino appeso alla finestra e i geranii sul davanzale. Le abbisogna la libertà della sua camera – è stanca, si butta a giacere – fruga negli armadi – contempla le sue tendine stirate di fresco – passa in rassegna i suoi fazzoletti. Trova un giubboncino smesso: « a chi darlo? ah ! sì a Carletto. Come sono contenta! » ha voglia di cantare? – è inutile bisogna che canti, altrimenti le piglia il nervoso. Ma la vicina di sopra è ammalata: – « che cervellino! Non canto più – vado a trovarla, le ho promesso alcuni biscotti. »

Eccola che sale come una cervetta – il suo cuore batte – la sua immaginazione è in moto – i suoi nervi sono gradevolmente tesi.

Arriva – abbraccia la vicina, le accomoda i guanciali, le offre da bere – gira per la camera, osserva, si dà da fare. In un'ora il suo cervello ha riflesso cento immagini, ha avuto cento pensieri; la sua fantasia lavora, lavora – e cambia sempre. Ecco la donna.

Una creatura così mobile, sensitiva, impressionabile – cangiante come l'iride – viva come la polvere – oh! perchè vorreste inchiodarla otto ore di seguito su una orrenda sedia burocratica – dura, nera, lucida, colle borchie d'ottone e la spalliera di legno? Che cuore avreste di costringere i suoi occhi su un registro di scrittura doppia e obbligare la sua testolina a mettere in fila l'esercito ribelle dei numeri?

Il minore de' suoi mali sarà l'emicrania. È una bella giornata di primavera – ella aspira al sole, ai fiori, all'aria tiepida – i suoi polmoni hanno bisogno di dilatarsi – il suo petto delicato vuole respirare – una passeggiata le farebbe tanto bene! – No, somme e moltipliche.

È inverno – piove, nevica, fa un freddo indiavolato. Il suo corpicino si raggomitola voluttuosamente sotto le coltri – ancora dieci minuti .... dieci minuti appena! No – alzati, va fuori, cammina nella neve e nel fango, pigliati una infreddatura, una bronchite – lo studio ti aspetta.

Si alza.

« Lasciatemi qui in casa, al caldo, nella mia poltroncina; lavorerò molto, cucirò camicie, ricamerò, taglierò vesti. »

No – allo studio. Devi essere una donna indipendente.

Dunque la donna non deve far nulla?

Nulla? Se è nulla tutto quello che fece da Eva in poi, benedetto ozio!

Ella bada alle faccende domestiche, sorveglia i servi, fila lana e tesse tuniche per il marito, cresce i figli, è esempio ai vicini, specchio ed ornamento alla casa di suo marito.

Così la Bibbia.

Capisco (interrompe qui Gianni, un ortolano di mia conoscenza): sono belle parole, ma per noi poveretti che non possiamo tenere in casa nè specchi nè ornamenti, ci è ben più cara una donna che sappia adoperare una zappa e la vanga.

Caro Gianni, di voi e delle vostre donne parlerò un'altra volta; adesso mi occupo di quella classe che voi chiamate mezze signore, che hanno cioè l'educazione senza le rendite.

Tornate pure ai vostri piselli.

Dopo otto ore di lavoro come sperate voi di trovare la donna? Magari avrete la pretesa di vederla fresca, allegra, carezzevole, volteggiarvi intorno, apparecchiare la tavola, fare il dovere ai bimbi, ricucirvi il soprabito, sorridente come nulla fosse. Poi alla sera voi andate al caffè o all'osteria; ed ella giù da capo a lavorare cogli abiti, colla biancheria, coi mobili di casa.

Certe persone si mettono in mente di tenere que' begli uccelli d'America dalle penne variopinte, leggiadri, graziosi – verissimo, – ma che richiedono immense cure. Le persone ch'io dico non ne fanno caso; oggi l'uccellino perde una penna – domani ripiega l'ale – doman l'altro muore.

Così è la donna.

Vi lagnate che la vera donna è rara; perché dunque distruggerne la specie con istituzioni barbare che togliendo la donna al suo elemento di pace e d'amore ne fanno un essere impossibile, zotico, che a voi stessi non piacerà più quando la vedrete tornare dallo studio imbronciata, nervosa, piena di stizza, lorda d'inchiostro, colla penna sull'orecchio? .....

Siete poeta? Come troverete la vostra casa allorchè, rientrando, non vi sarà la moglie sorridente e lieta che vi accoglie, e con quella spigliata volubilità tutta propria del suo sesso vi narra i mille piccoli incidenti della giornata! E le sue idee fresche e graziose scendono come raggio di sole a dissipare le nebbie del vostro capo oppresso. Anch'ella è oppressa; ha la testa piena di numeri – cosa deve dire? – Via, parla, raccontami qualche novità, supplica il marito.

– Novità? Oggi abbiamo avuto il bilancio; le entrate superano di poco le uscite. Se non viene questo benedetto pareggio ....

Auf!!! –

Amate la prosa? Vi preme un bel pranzetto ben cucinato, fatto con gusto? Non potete avere un cuoco, questo si sa, ma dite alla moglie:

– Mia cara, oggi lo stufato non si poteva mangiare; ci ho trovato per mio conto una mezza dozzina di chiodi di garofano.

– Che vuoi! Sono all'ufficio tutto il giorno, – risponde giudiziosamente la moglie.

Michelet[14] ha scritto che la donna è un'inferma.

Grazie al cielo non è vero: il signor Michelet esagera in questa come in moltissime altre cose. Ma è pur vero che il fisico della donna va soggetto a tali e tante variazioni da non potersi dire ch'ella sia sempre in salute.

Fanciulla ha l'isterismo – sposina, ha i vapori – madre .... – Qui vi vorrei vedere i signori uomini; per quanto ne parliate, siete giudici incompetenti. Mettetemi un po' una donna in quello stato per tutto il giorno dietro un tavolo, curva sul mastro!

E i bambini chi li cura? Chi li avvezza? Chi li educa?

Vi sono gli asili. Ma in nome di Dio, se mi cacciate la donna in uno studio, i figli all'asilo, date pur fuoco alla casa e che non se ne parli più.

Tant'è tanto, per mangiare e per dormire c'è l'osteria.

Il progresso e la civiltà, una bella coppia che ha dato al mondo bellissimi figli, ne ha però contro sua voglia partorito anche de' bruttini – qualcuno losco, qualch'altro zoppo. – Si sa, la perfezione non è nella natura umana. Tra questi brutti figlioli ve n'ha uno che s'è accasato male ed ebbe una prole numerosissima di *spostati*.

Gli spostati sono proprio una creazione del nostro secolo ed aumentano ogni giorno – invadono le capitali, prendono d'assalto le persone benefiche e i posti vacanti. S'io volessi ora descriverne la genealogia, perderei troppo tempo, allontanandomi dal soggetto principale del mio discorso – ma se lo bramate vi tornerò sopra.

Per il momento la mia intenzione è di farvi conoscere che togliendo la donna alla casa e lanciandola negli uffici pubblici non farete che accrescere il numero degli spostati. Ideale e materia, gene ed economia, tutto tutto si accorda per lasciare la donna nel posto che ha finora occupato di figlia, di sposa, di madre, di educatrice, di infermiera, di consolatrice e ispiratrice – non mai di scrivana e di contabile!

Quello che mi dà fiducia di persuadere i lettori è che io non scrivo per spirito di parte;

– la quistione, personalmente, non mi interessa affatto. Non sono fanatica né per il sesso debole né per il sesso forte – amo la verità – ed amo anche la libertà per tutti.

Vorrei che si lasciasse libera agli uomini la carriera degli impieghi e che per una illusoria prospettiva di lucro non si trascinasse la donna a rinunciare al suo massimo bene che è la casa e la famiglia, con tutti i suoi piccoli bisogni, co' suoi lavori tranquilli, colla sua poesia, colla sua pace.

All'aquila le rupi, al leone la foresta, alla tigre il deserto: – agli uccelletti e alla donna, il nido.

Neera

L' Illustrazione Italiana, III. 25 (16 aprile 1876)

La Donna Povera

Lettera della Marchesa Colombi alla signora Neera

Senta, signora Neera; io le giuro sulla testa calva della destra parlamentare, sulla testa arruffata della sinistra, sulla testa grigia del centro, che non aspiro a vedere la donna in Parlamento. – L'assicuro che, quando lessi nei giornali di moda che si usavano le corazze per le signore, ho avuto un momento di vero terrore all'idea che, dietro le fisime dell'emancipazione della donna, fosse sorto un reggimento di corazzieri in gonnella, armati della *spada di Damocle*, dell'*usbergo del sentirsi puri*, e di tutti gli altri amminicoli dell'armeria rettorica, per movere, sotto la scorta dell'onorevole Morelli,[15] all'assalto di Montecitorio. – Il giorno in cui dovessi vedere co' miei propri occhi una deputatessa, una medichessa, un'avocatessa sotto il bel cielo d'Italia, sarebbe il giorno più triste della mia vita. Questo per provarle che, fino ad un certo punto, siamo d'accordo. – Ma, tra i deliri degli emancipatori, ed il beato *Domum servavit et lanam fecit*, da cui ella non vuole che la donna si diparta, c'è un abisso.

C'è uno di quei figli del progresso e della civiltà che ella deplora. – Bruttini, loschi, zoppi quanto vuole. Ma esistono; e dacché non possiamo gettarli nel Taigeto[16] e farla finita, bisogna pure che ci rassegniamo agli incomodi rimedi dell'ortopedico.

Questo figlio zoppo è il ceto civile e povero; è la grande famiglia delle *mezze signore* di cui ella parlava nel suo brillante articolo nel N.23 dell'ILLUSTRAZIONE.

Per queste mezze signore, se non trovo un vantaggio nelle ampollose promesse degli emancipatori, non lo trovo neppure nella poesia delle idee da lei espresse, e sono costretta ad esclamare come il mio povero marito buon anima:

– Tra l'uno e l'altro son di parer contrario.

È verissimo quanto ella dice che *la donna è nata per piacere agi uomini, propagarne la specie, migliorarla, ingentilirla e far calze.*

In principio sono perfettamente del suo parere. – Se una donna ha da vivere, se la necessità non la spinge a guadagnarsi l'esistenza, il posto della donna è la sua famiglia e non deve uscire di là. – I pubblici di tutti i teatri d'Italia possono far fede della mia ripetuta ed esplicita dichiarazione, che:

«Il cor mi balza
Di domestica gioia lavorando la calza.»

Ma pur troppo vi sono delle eccezioni: la numerosa prole di quel figlio zoppo. Ne scelgo due nel suo stesso articolo: La fanciulla senza dote, e il giovane impiegato a mille e dugento lire.

– Lascio la situazione come l'ha posta lei.

«Guadagna mille e dugento lire. Ha conosciuto una donna. – Riamato l'ama. – Si parlano, si scrivono, sospirano – a quando le nozze? Ma ! – Il principale non vuole crescere lo sipendio, gli impegni di una famiglia sono molti. – Se venisse un terno al lotto! – Il terno non viene.»

Questa situazione ella la scioglie così.

«La fanciulla, cui fuggono gli anni, sposa un altro. – Egli resta solo, disilluso, infelice.»

Ma chi, *un altro*? Un ricco, no; perché i ricchi che sposano ragazze senza dote sono mosche bianche. – E le mosche bianche non fanno regola. – Dunque – un altro povero, su per giù, come quello che lascia? Allora tanto faceva che lo sposasse lui. E ad ogni modo, quegli sposi saranno sempre nel caso di tirarle verdi insieme finché dura l'amore, e di pianger divisi sul morto quando la miseria lo avrà ucciso. E se poi non lo sposa, un altro? – Se affezionata a quell'uno non può dimenticarlo? – Le donne non hanno la forza fisica dell'uomo – non hanno la sua energia di carattere; né la sua serietà; né la sua intelligenza. Questo lo dicono tutti; sarà vero. Ma nemmeno i più accaniti detrattori delle donne hanno mai conteso loro il sentimento. E vorrebbe negarlo lei, signora Neera?

Lei che è una donna, deve sapere come amino le donne!

Mi par di udirla rispondermi, con quel suo brio che mi piace tanto: –

«Ma, signora mia, una donna ben fatta e spiritosa è più facile che sappia come amano gli uomini.»

Ebbene, lo dica; ed avrà ragione lei, perché chi ci fa ridere ha sempre ragione: – Ma per ora mi faccia supporre che quella ragazza ami abbastanza il povero impiegato da non poterne sposare un altro. Ed in questo caso, – che è il più frequente, – resta anch'essa sola, disillusa, infelice come lui.

E l'uomo ha altri pensieri che lo aiutano a dimenticare le sue miserie personali. La politica è una preoccupazione vivissima, che appassiona l'uomo, fino a scindere le amicizie più care, fino al sacrificio. – Leggero di borsa, ed anche di stomaco, un uomo si riscalda sopra un articolo di legge; si accapiglia in un *meeting* elettorale. – E nei giorni di crisi politica prende il fucile, e via! Camminerebbe sul cadavere de' suoi cari per correre a battersi pel suo paese. – Parlo degli uomini ben organizzati, dei veri uomini, veda. Perché quello là del suo articolo che invidia alla sorella la libertà di stare a casa, quello non vale la pena che ce ne occupiamo.

La povera fanciulla invece è molto indifferente al colore dei candidati che dovranno andare a Montecitorio. Le elezioni non la distraggono dal suo dolore, non animano il suo isolamento. – La guerra la fa penare per sé e per gli altri, e la rende maggiormente infelice.

Intanto gli anni passano. – Il giovane ha fatto qualche passo nella sua carriera; vive modestamente, ma vive del suo lavoro; è un uomo. Ma la fanciulla non è una donna; – è una zitellona; quella specie di essere ibrido che non sa né come annunciarsi, né come vestire, che non appartiene a nessuno, che vive presso uno zio, presso un parente qualunque; – che non ha casa. Ha perduta la speranza, non ha preoccupazioni serie a cui volgere la mente; pensa ancora coi capelli grigi al suo amore svanito, ed il mondo ride della vecchia che passa la sua vita a tirar un idillio per la coda.

E, se ha l'animo delicato si sente umiliata della sua posizione. Vorrebbe non essere a carico di nessuno; avere la sua casa; bastare a se stessa. Ma come fare? Non ha denaro. Non sa guadagnarsene.

E ce ne sono tante, signora mia, che hanno mancato al primo de' suoi precetti: *Piacere agli uomini.* E i suoi precetti sono come gli anelli di una catena. Uno tira l'altro. Se non si piace ad un uomo, non si può propagarne la specie, né altro. Resta la risorsa dell'ultimo precetto: Far calze, a una lira e cinquanta centesimi al paio; e se ne possono fare due paia alla settimana!

Se invece quella fanciulla si fosse innalzata fino a certe alte missioni del sesso forte ed intelligente, – di legger le soprascritte ed imprimere bolli alle lettere negli uffici di posta, o di distribuire biglietti di strada ferrata, o di trasmettere dispacci telegrafici, o di notare nel libro maestro d'un negozio di mode le trine della signora tale, i fiori della signora tal altra, – rimanendo zitella avrebbe avuto di che tenersi la sua casetta, di che bastare a sé stessa, di che occupare la sua attività ed il suo spirito intelligente.

Aggiunga che quel giovane povero avrebbe potuto sposarla, ed unendo il frutto delle fatiche comuni avrebbero potuto tirare innanzi, finché l'aumento di stipendio del marito gli permettesse di mettere la moglie al dolce riposo della vita domestica.

A questo ella mi risponde: «*Che marito sarà quello che specula sul lavoro della moglie?*»

Che marito? Ma un marito come tutti gli altri, che speculano, – dacchè vuol dire questa parola, – sulla dote delle loro mogli.

Come? Il marchese Alfonso, che apre il suo scrigno a due battenti per farvi entrare le ventimila lire di rendita della signorina Matilde, sarà un fior di galantuomo, ed un gentiluomo per soprammercato. – e quel poveretto che, per non piantar in asso una povera ragazza che lo ama, si rassegna a lasciarla lavorare per contribuire al mantenimento della famiglia, sarà una canaglia?

Scusi, signora Neera, ma io non l'intendo.

– Aiutarsi scambievolmente, non fu sempre il senso morale del matrimonio?

– Io ammetto che la donna sta meglio come ella descrive: – «accanto al focolare colle sue gaie faville, coll' amico caffè che gorgoglia, colla comoda seggioletta a bracciuoli, ed il canarino, e i geranii, e nella libertà della sua stanza, dove si butta a giacere, fruga negli armadii, contempla le sue tendine stirate di fresco, passa in rassegna i suoi fazzoletti» e visita le vicine malate, e dona le robe smesse in elemosina, e cucina lo stufato pel pranzo.

Ma cogli stipendi che hanno molti mariti delle *mezze signore*, una donna che volesse viver così, correrebbe il rischio di non avere né faville sul focolare, né caffè nella cogoma; e frugando negli armadii, potrebbe contare i suoi fazzoletti e le sue tendine sulle ricevute del Monte di Pietà. – E lo stufato poi! Per un impiegato a cento lire al mese, colla moglie da mantenere e vestire, e la pigione da pagare, e la ricchezza mobile, e il suo sarto, e il calzolaio, ed un cappello al mese che sciupa in saluti ai superiori.... ah, se la moglie non ci rimette del suo, la minestra è già un problema difficile da risolversi ogni giorno; – e lo stufato poi! ah, lo stufato è un sogno, un'aspirazione, un poema!

Io, vede, ho un'opinione esaltata addirittura, del senso estetico di tutti i Travetti[17] del regno – Ma tuttavia ho un lontano sospetto, che, – posti nell'alternativa preferirebbero veder la moglie come la dipinge lei tornar dal lavoro « zotica, imbronciata, nervosa, piena di stizza lorda d'inchiostro, colla penna sull'orecchio »; ma che avesse dato alla mattina ordini e denari ad una serva per un modesto pranzetto, – che trovare la sposa a far la calza sorridendo, – ed il gatto sul fuoco.

Del resto, lo crede proprio lei, signora Neera, che una donna per stare ad uno scrittoio debba ridursi così trista e brutta? Badi; ella scrive bene come pochi uomini sanno scrivere. È laboriosa, e deve starci parecchio tempo al suo scrittoio. – Ebbene, guardi: Io le giuro sul capo innocente del suo « Carlotto in città »[18] che me la figuro elegante, graziosa; che desidero vivamente di conoscerla; e non potrei mai e poi mai immaginarmela *piena di stizza, lorda d'inchiostro e colla penna sull'orecchio.* È impossibile. – ha troppo garbo, troppo spirito, è troppo donna perchè uno scrittoio possa deformarla così.

E anche senza il garbo e lo spirito che ha lei, io conosco delle signore che fanno le maestre di pianoforte e di lingue straniere, che corrono tutto il giorno in omnibus, e su e giù dalle scale, qualunque tempo faccia, a dar lezioni in una casa e nell'altra, e non sono meno amabili e meno signore per ciò. E ne conosco altre che sono direttrici e maestre di scuola, e stanno otto ore di seguito sopra una cattedra, e non hanno solo « *un esercito di numeri* » da mettere in fila in un registro, ma un esercito di bambini chiassosi ed irrequieti da tener a segno. Ed oltre ad occupar la mente per istruirli debbono parlare per ore ed ore, gridare, sciuparsi i polmoni. E l'aria d'una scuola, in cui respirano sessanta bambini, non tutti puliti, coi profumi delle pere fracide e delle croste di formaggio che si nascondono in tasca, non è più sana di quella d'uno studio dove stanno degli impiegati. – Eppure quelle signore ci vivono, e lo fanno volentieri per il bene della loro famiglia, e, coll'energia che dà l'amore, sanno cogliere le ore del mattino e della sera per occuparsi della loro casa.

O perché chi s'oppone a che la donna faccia concorrenza all'uomo in certi facili impieghi, non mette poi nessun ostacolo, anzi trova naturale, che faccia la maestra? Perché è più faticoso e rende meno, forse? – Sarebbe generoso.

Ancora una volta, se l'uomo lo può lasci alla moglie il solo lieve incarico della sua famiglia e la renderà più felice, e la manterrà più bella, e sarà più d'accordo con la società. – Ma quando per tenerla in casa disoccupata è costretto a condannare sé e lei a mille privazioni, – santa la donna che osa gettarsi coraggiosamente al lavoro.

I tempi sono mutati; il progresso e la civiltà hanno creato il figlio zoppo. – Non cerchiamo di scoraggiare la donna povera dal lavoro. È il rimedio dell'ortopedico.

La Marchesa Colombi

La Lega della Democrazia,[19] I. 132

(15 maggio 1880)

Lettera aperta di Anna Maria Mozzoni a Matilde Serao

In fatto di politica, le donne non hanno una opinione, hanno un sentimento. Quindi amano la monarchia. È la politica del cuore, ad è anche la mia.

Matilde Serao

È egli permesso, amabile Signora, di intervenire nella polemica che si dibatte fra lei e il dottor Bertani all'ombra del diritto di difesa?

Ella afferma che le donne non hanno opinione in politica e che esse amano la monarchia perché è il partito del cuore, ed evidentemente ella intende di fare alle donne un elogio ed io, guardi! ho la malinconia di riguardare la sua prima affermazione come un insulto e la seconda come una stranezza.

E dire che queste due affermazioni così poco vere sono della brava autrice degli studi "Dal Vero"?![20]

Possibile, Signora mia, che ella non abbia mai conosciuto donne che sentano di appartenere all'umanità e alla nazione e si diportino in conseguenza? Debbo credere che il cuore sia proprio incaricato di intenerirsi sul conto dei gaudenti anziché dei sofferenti e di commuoversi d'affetto per chi nuota negli agi e nella grandezza, anziché d'affettuosa pietà pei molti che affogano nelle strette del bisogno e dell'abiezione? Che cuore curioso!

Eppure, Signora, veda la differenza che corre da un cuore all'altro e come per esempio, io senta altrimenti da lei pur essendo una donna.

La mia ragione era acerba come l'età: ed un giorno, che non dimenticherò mai, le vie di Milano erano deserte, chiuse le botteghe e le porte: il popolo si affollava muto e pensoso sulla piazza del castello dove stavano rizzati sette patiboli. Dei lugubri rintocchi suonavano l'agonia di sette uomini che camminavano verso quelle forche, legati, fra due schiere di soldati, seguiti dal carnefice, accompagnati dal rullo intermittente dei tamburi velati a bruno. – Quegli uomini espiavano l'amore della patria e non avendo potuto liberarla col braccio ne fecondavano la libertà col sangue.

Era fra quegli Antonio Sciesa, un cartolaio, Signora mia, né più né meno. Un ufficiale gli si accostò e gli disse "Sciesa, siete ancora in tempo; svelate i vostri complici e avete la vita e la libertà". – Sciesa, il cartolaio, Signora, né più né meno, si volse al carnefice e gli disse con tutta semplicità: « Tira innanzi! » e Sciesa salì al patibolo senza svelare i complici, senza ammirarsi, senza arringare il popolo, senza forse neppur sospettare che i nove decimi del genere umano passato e presente non lo valevano.

Ebbene che vuole, cara Signora, io ero giovanissima, ma quel cartolaio sconvolse tutte le idee che una educazione ultra clericale mi aveva piantate nella mente e quelle due parole mi parvero più monumentali che le sette parole del giusto crocifisso. Da quel giorno io ebbi convinzioni in politica, benché donna, e forse perché donna non si svigorirono più. Seppi poi di Ciro, di Alessandro, di Carlo quinto, di Enrico quarto, di Napoleone primo e di altri, tutti principi valorosissimi, tutti eroi di primo rango; ma io avevo fatto nella mia mente due categorie di eroi e stetti fissa alle mie categorie, l'una che fu chiasso e si glorifica nel sangue altrui, l'altra che non fa rumore, e che prodiga il proprio – e vede un po', quando si dice, la qualità del cuore! – io ho sempre optato per la seconda.

Né le acclamazioni della piazza mi impressionarono più che le insinuazioni della storia. Vidi il popolo affollarsi sul passaggio della bellissima imperatrice d'Austria ed acclamare a quel viso di donna prodigiosamente geniale del quale gli occhi non si saziavano – vidi in un nido di raso o di cristalli la bella ed infelice principessa, che fu poi imperatrice del Messico, e che giungeva fra noi preceduta da una fama di bontà e di intelligenza che la sua condotta non ismentì – le mani plaudivano – le grida salivano al cielo – il mio cuore non si commosse. Vidi Napoleone III e Vittorio Emanuele entrare sotto l'arco di trionfo, dopo Solferino – amai figurarmi che in quei due petti riscaldati da sangue italiano annidasse un sentimento che non fosse esclusivamente la soddisfazione di una speculazione politica riuscita a personale vantaggio. Ma nessun luccicare di metalli, nessun ondeggiar di piume, nessuna acclamazione di popolo, nessuna bellezza di volto gentile, o marziale, valse mai a far prevalere nell'animo mio la grandezza del rango o del successo, alla grandezza della virtù oscura ed infelice – e né enciclica di papa, né proclama di generale vittorioso, né discorso del trono, né dialettica d'oratore mi parve mai bella, magnanima, eloquente e convincente quanto quel "Tira innanzi" del povero cartolaio.

Dopo ciò, mia cara Signora, pensi s'io sia disposta ad ammettere che la monarchia sia il partito del cuore. Che si possa essere monarchici ed avere ogni pregio d'anima o di mente non vi è chi non ammetta ed ella ne è valido documento; ma che la monarchia ed il cuore siano proprio cugini germani, il buon senso non lo può ingollare e protesto. Auguro anzi a questo partito di contare fra i suoi aderenti delle donne che non siano esclusivamente partigiane delle feste, dei balli, dei ricevimenti e delle brillanti *toilettes*, del che ne sospetto più d'una; e desidero di cuore alla monarchia, perché adoro la virtù ed il carattere, anche una sola donna dello stampo della principessa di Lamballe.[21] Ma non lo spero troppo, perché la monarchia, non tanto per sua colpa, quanto per la natura sua, ha sempre fatto dei beati che scappano prima di diventar martiri – la storia della rivoluzione informi.

Venendo poi, cara signora, all'altra affermazione sua che le donne non hanno opinione in politica, io le risponderò, non affermando ma provando. Si può perdonare ad una donna colta com'ella è d'ignorare quale e quanta parte abbiano preso le donne nelle questioni che hanno agitato i loro rispettivi paesi in tutto il mondo civile? Quale lotta e quanto intelligente sostengano in America ed in Inghliterra pel loro voto politico? Può ella ignorare che le francesi s'interessano tanto alla politica che cercano l'ammissione agli Stati Generali fin dal 1789, e con tali ragioni che se oggi fanno ridere i fanfullisti convinsero allora Sieyès, Marat, Danton, Garlier e Condorcet? e tutti gli sforzi ch'esse ripeterono da allora fino ad oggi inclusivamente per ottenere il voto? Può ella ignorare la larga parte, anzi le iniziative che si devono alle donne nella abolizione della schiavitù, nella riforma penitenziaria, nel richiamo degli atti su malattie contagiose e nella riforma delle loro condizioni giuridiche e sociali? Può ella ignorare che le donne collaborarono al movimento filosofico che preparò la rivoluzione, che si trovarono in tutte le fasi della rivoluzione stessa, che si videro nella restaurazione, e che in tutti i partiti con la penna, con l'opera, coi sagrifizi e col carattere si mostrarono sempre ad un'altezza che fermò l'attenzione di Jourdan, di Michelet, di Lairtuillier, di Fourier e persino di Prondhon, loro sistematico dispregiatore? Può ella ignorare che queste donne ch'ella afferma non aver opinioni politiche, scrivono giornali, taluni diffusissimi ed eccellenti in America, in Inghilterra, in Francia, in Italia, in Germania e vi discutono tutte le questioni e rivendicano il diritto di voto? Può ella ignorare le gesta delle nihiliste e l'eroismo, l'entusiasmo, la fede e l'energia che esse profondono al servizio della causa che hanno abbracciata?

Che le donne non hanno opinioni politiche bisogna lasciarlo dire agli avversari del loro voto elettorale perché nella controversia la buona fede è rara e la lealtà è troppo spesso un desiderio – ma a lei, signora mia, come mai non le è giunta ancora la notizia che, grazie a Dio, il genere femminino possiede delle varietà non meno meritevoli delle eleganti, e non sempre belle, ignoranti che ella considera come prototipi del sesso?

E dacché ho cominciato mi consenta, cara Signora, ch'io vuoti il sacco tutto in una volta e le dica aperto l'animo mio che, tanto e tanto, Ella ha troppo ingegno per volermene male, e quanto a me, sbarazzata da quell'uggia che mi fastidisce, non vi penserò più.

Io posseggo i di lei studii *Dal Vero* che ho addittati io stessa all'attenzione del signor Bertani, pel gusto squisito della forma o la grazia efficace dei concetti. Non conoscendola personalmente attraverso a quel volumetto io le voleva bene. Solo mi spiacque quel deprezzamento del suo sesso che vi si legge fra le righe, e trabocca tutto quanto laddove fingendo alle donne l'esercizio del voto si compiace nel figurarsele assorbite da puerili e ridicole brighe e mosse da piccoli e indegni moventi, – Ebbene, ma lo creda, quel bozzetto, malgrado il merito della forma, non merita posto in un volume che s'intitola *Dal Vero* prima perché le donne non avendo il voto fra noi, ella non ha cavato *dal vero* ma *dal presunto* – secondo, perché dove le donne esercitano il voto lo fanno in un modo tanto diverso da quello da lei supposto e così superiore ad ogni malevolezza da fare ben torto ad una scrittrice colta come lei di ignorarlo e ad ogni persona intelligente di non apprezzarlo.

Quando gli uomini si prendono lo spasso di berteggiare le donne, siano essi mossi da gelosie d'impiego come il dottore Onesti, o dal dovere di far ridere ogni giorno di qualche cosa come *Fanfulla*, si capisce e non ci si abbada – ma quando una donna si incarica di diminuire il suo sesso, contro la verità, vi travedo una tacita adulazione all'altro che mi pare ignobile e abietta come tutte le adulazioni.

Non mi voglia male, simpatica Signora, se mi permetto di darle un consiglio, non già perché me ne supponga un'ombra di diritto, ma per compensarmi del dispiacere delle sue affermazioni sul conto di tutte le donne – eppoi, che serve? – gliela dico tutta perché mi pare che il di lei ingegno abbia un fondo più consistente di quello che i suoi primi scritti, per quanto lodevolissimi, sembrano annunciare.

Dilati i suoi orizzonti e spinga lo sguardo al di là delle sale e dei palchetti e troverà donne diverse da quelle che dopo aver giocato da bimbe a far la signora, diventate signore giocano a far la bimba, affettando ignoranza e incompetenza in tutte le cose, salvo poi a tenersi sul tavolino Lo *Scannatojo*[22] e la *Nanà*[23] ed a mandare a memoria versi di Lorenzo Stecchetti.[24]

Ma al di là di queste ve ne sono altre e diverse.

Pur troppo non è convenzionale or basato sui fatti quello che il signor Bertani asseriva delle inglesi, delle russe, delle americane. Legga i loro libri ed i loro giornali, Signora mia, e si informi degli oggetti della loro attività, eppoi da quel franco carattere ch'ella è, dovrà dire, pur troppo la nostra cultura è nulla, la nostra educazione è autoritaria, la nostra letteratura è di mussola e i nostri caratteri riescono di garza maria.

Oh se le italiane che hanno ingegno e coltura, e non son poche, si facessero capaci di questo anziché disconoscere irriflessivamente la attuale superiorità delle donne di altri paesi si deciderebbero ad uscire una buona volta d'infanzia, smetterebbero quella incompetenza universale e convenzionale, si mostrerebbero quali sono intelligentissime o si formerebbero dei caratteri morali pari all'intelligenza.

Mi creda con tutta stima, cara Signora,
Sua devotissima.

La Lega della Democrazia, I. 138 (21 maggio 1880)

Lettera aperta a Anna Maria Mozzoni di Matilde Serao

La signorina c'invia la seguente lettera

Pregiatissima Signora Mozzoni,
Sulla questione: *politica del cuore*[25] ebbi una cortese polemica artistica nel *Corriere del Mattino*, col valente scrittore e direttore di quel giornale Martino Cafiero. Le invio i due numeri dove si contiene la polemica, per non dover ripetere quello che ho già scritto.

Per quello che riguarda la donna e la politica, non posso discutere con lei, pregiata signora. Se me lo concede, io sono un'artista. Naturalmente i democratici, i repubblicani, emancipatori, socialisti, umanitarii, signori e signore – non son artisti. Ditalchè mi trovo costretta a dirle che quella certa specie di donna che è l'ideale delle sue teorie, è da noi assolutamente respinta, come una figura rigida, dura, senza alcuna poesia. Anzi, per riunire tutti questi aggettivi, noi la respingiamo come antiestetica. Per la vita, per l'amore, per l'arte ci vuole la donna. Istruita, ma donna. Educata, ma donna. Maestra del popolo, infermiera, scrittrice, educatrice, ma donna. Niente diritti politici, niente ingerenze elettorali, niente attribuzioni maschili, niente professioni impossibili. La donna, egregia signora Mozzoni, stupenda creazione che le teorie dei democratici, repubblicani, emancipatori, socialisti, umanitarii, signori e signore, non arriveranno mai a demolire.

Sebbene a lei possa parere impossibile, ho letta la storia. Salvo le debite e nobili eccezioni, so di donne più o meno eroine, dalle virtù più o meno equivoche, dalle reputazioni più o meno usurpate. È il caso di chiedersi se queste donne avevano uno sposo, se lo amavano, se avevano figli, se sapevano insegnar loro l'alfabeto, se i mobili delle loro case erano spolverati a dovere, se sapevano cucire, rimendare, far di conti, – cose molto importanti. Per semplice curiosità, senza nessun interesse, senza nessuna ammirazione leggo delle socialiste e delle nihiliste. Sono casi patologici, deviazioni del cervello che soffoca il cuore, aberrazioni del senso femminile. La scienza li studia. Per fortuna la donna italiana, come sempre, dà esempio di moralità e di buon senso.

Io accetto il suo consiglio di dilatare il mio orizzonte. Quando i miei lavori artistici me lo permetteranno, io mi propongo di combattere in tutti i modi, ad oltranza, senza remissione, specialmente con l'arma del ridicolo che è in questo caso la migliore, la falsa emancipazione, i diritti politici inutili e dannosi, la letteratura ampollosamente emancipatrice, i giornaletti femminili, clandestini ed emancipatori, la donna-apostolo; la donna-missione, la donna-principio. Questo per sentimento di scrittrice, per coscienza della società femminile.

Sono dolente che fra i miei sentimenti e le sue teorie, egregia signora Mozzoni, il dissenso sia così profondo. A stare tutta la nostra vita insieme, non saremmo mai d'accordo. Sono anzi tanto certa della mia impertinenza, che essendo stata obbligata a dichiarare ampiamente quello che sento, non aggiungerò una parola di più, se democratici, repubblicani, emancipatori, socialisti, umanitarii – signori e signore – vorranno continuare ad occuparsi di me.

Con tutta stima, signora, mi creda.

Napoli, 10 maggio 1880

Devotissima

Matilde Serao.

Notes

1. Postal Culture after 1861

1 Emilia Toscanelli Peruzzi, *Diario*. Ed. Elisabetta Benucci (Florence: Società editrice fiorentina, 2007), 61. Unless otherwise noted, all translations from the Italian are mine, with the help of Alyda Stabile.

2 Edmondo De Amicis, *La lettera anonima* (Genoa: ECIG, 1991), 52–3.

3 Given the vast bibliography available on this topic, I will mention only critical works published since the 1990s: Thomas O. Beebee, *Epistolary Fiction in Europe: 1500–1850* (Cambridge: Cambridge UP, 1999) and "Publicity, Privacy, and the Power of Fiction in the Gunning Letters," *Eighteenth-Century Fiction* 20, 1 (2007): 61–88; Roger Chartier, Alain Boureau, and Cécile Dauphin, eds, *Correspondence: Models of Letter Writing from the Middle Ages to the Nineteenth Century* (Princeton: Princeton University Press, 1997); Adriana Chemello, *Alla lettera: Teorie e pratiche epistolari dai Greci al Novocento* (Padua: Guerini Studio, 1998); Jane Couchman and Ann Crabb, eds, *Women's Letters across Europe: Form and Persuasion* (Burlington, VT, and Aldershot, Hants: Ashgate, 2005); James Daybell, ed., *Early Modern Women's Letter Writing* (Basingstoke: Palgrave, 2001); William Merrill Decker, *Epistolary Practices: Letter Writing in America before Telecommunications* (Chapel Hill and London: U of North Carolina P, 1998); Margherita di Fazio, *La lettera e il romanzo: Esempi di comunicazione epistolare nella narrativa* (Rome: Nuova Amica Editrice, 1996); Mary A. Favret, *Romantic Correspondence: Women, Politics and the Fiction of Letters* (Cambridge: Cambridge UP, 1993); David Henkin, *The Postal Age: The Emergence of Modern Communications in Nineteenth-Century America* (Chicago: U of Chicago P, 2006); James How, *Epistolary Spaces: English Letter Writing from the Foundation of the Post Office to Richardson's Clarissa* (Aldershot, Hants, and Burlington, VT: Ashgate, 2003); Katharine Ann Jensen, *Writing Love: Letters, Women, and*

the Novel in France, 1605–1776 (Carbondale and Edwardsville: Southern Illinois UP, 1995); Linda Kauffman, *Special Delivery: Epistolary Modes in Modern Fiction* (Chicago and London: U of Chicago P, 1992); Elizabeth J. MacArthur, *Extravagant Narratives: Closure and Dynamics in the Epistolary Form* (Princeton, NJ: Princeton UP, 1990); Richard Menke, *Telegraphic Realism: Victorian Fiction and Other Information Systems* (Stanford, CA: Stanford UP, 2008); Esther Milne, *Letters, Postcards, Email: Technologies of Presence* (New York: Routledge, 2010); Armando Petrucci, *Scrivere lettere: Una storia plurimillenaria* (Rome-Bari: Laterza, 2008); John Richard, *Spreading the News: The American Postal System from Franklin to Morse* (Cambridge, MA: Harvard UP, 1995); Bernhard Siegert, *Relays: Literature as an Epoch of the Postal System* (Stanford, CA: Stanford UP, 1999); Gino Tellini, *Scrivere lettere: Tipologie epistolary nell'Ottocento italiano* (Rome: Bulzoni, 2002); Barbara Zaczek, *Censored Sentiment: Letters and Censorship in Epistolary Novels and Conduct Material* (Newark and London: U of Delaware P, 1997); Gabriella Zarri, ed., *Per lettera: La scrittura epistolare femminile tra archivio e tipografia secoli XV–XVII* (Rome: Viella, 1999).

4 On the transformation of the nineteenth-century print industry in Italy see Valerio Castronovo, "Cultura e sviluppo industriale," *Storia d'Italia* 4, ed. Corrado Vivanti (Turin: Einaudi, 1981), 1261–93; Fausto Colombo, *La cultura sottile: Media e industria cultural in Italia dall'Ottocento agli anni Novanta* (Milan: Bompiani, 1991); Giovanni Ragone, *Un secolo di libri. Storia dell'editoria in Italia dall'Unità al postmoderno* (Turin: Einaudi, 1999); Ann H. Caesar, Gabriella Romani, and Jennifer Burns, eds, *The Printed Media in Fin-de-Siècle Italy* (Oxford: Legenda, 2011).

5 Foscolo's epistolary novel has, for instance, been criticized for its lack of thematic and structural coherence and considered a product of the author's still premature poetic voice. See Walter Binni's introduction to *Ugo Foscolo, Ultime Lettere di Jacopo Ortis* (Milan: Garzanti, 1974), xvii–xl, and, more recently, Giuseppe Nicoletti's "Introduzione," in which he reiterated Binni's reading of the Ortis as "opera giovanile" and essentially incomplete, in Ugo Foscolo, *Ultime Lettere di Jacopo Ortis* (Florence: Giunti, 1997), xxix. Maria Antonietta Terzoli's *Foscolo,* additionally, highlights the derivative and therefore marginal aesthetic value of the *Ortis,* shaped thematically, stylistically, and linguistically by other European epistolary works as well as by Foscolo's own poetry (Rome-Bari: Laterza, 2000), 38. A different approach, however, may be found in critics such as Alberto Cadioli and Gino Tellini, who examined the *Ortis* within a context of Italian early-nineteenth-century theorizations on the novel. In particular, these critics' attention to the relationship between the *Ortis* and its readers showed the complexity of a novel that became the object of a real cult

of veneration for generations of Italians. See Alberto Cadioli, *La storia finta: Il romanzo e i suoi lettori nei dibattiti di primo Ottocento* (Milan: Il Saggiatore, 2001); and Gino Tellini, "Sul romanzo italiano di primo Ottocento: Foscolo e lo sperimentalismo degli anni Venti," *Studi Italiani* 13, 2 (1995): 47–97.

6 For Giusepp-Antonio Costantini see Cadioli, *La storia finta*, 34–6. Pietro Chiari wrote several books in the epistolary form: *La filosofessa italiana* (1753), *Pamela maritata* (1759), *Francese in Italia* (1759), *La viaggiatrice* (1760), *La cantatrice per disgrazia* (1762), and *La donna che non si trova* (1768). These texts reflect the influence of Richardson's *Pamela*, which was translated into Italian in 1744–6. See Arturo Graf, *L'Anglomania e l'influsso inglese in Italia nel secolo XVIII* (Turin: Ermanno Loescher, 1911), 282, and more generally on Chiari's novels, Carlo Alberto Madrignani, *All'origine del romanzo in Italia* (Naples: Liguori, 2000). Not only was the epistolary genre well known in Italy – as, in both the sixteenth and seventeenth centuries, several literary works were published in the letter form – but according to some critics it might have originated in Italy. See G.F. Singer, *The Epistolary Novel: Its Origin, Development, Decline and Residuary Influence* (New York: Russell & Russell, 1963), 189, and Beebee, *Epistolary Fiction in Europe*, 86. In 1569, Aloise Pasqualigo published *Lettere amorose*, while in 1684, Giovanni Paolo Marana wrote *L'esploratore turco e le di lui relazioni segrete alla Porta Ottomana scoperte in Parigi nel regno de Luigi il Grande*, which presumably influenced the writing of Montesquieu's *Lettres persanes* (1721). See Singer, *The Epistolary Novel*, 189–90, and Beebee, *Epistolary Fiction in Europe*, 86. The appearance of the epistolary novel certainly followed the popularity of books of letters (*lettere familiari*). Between 1538 and 1627, 540 volumes of letters were published in Italy. See Amedeo Quondam, ed., *Le "carte messaggere." Retorica e modelli di comunicazione epistolare: Per un indice dei libri di lettere del Cinquecento* (Rome: Bulzoni, 1981), 30; Maria Luisa Doglio, *Lettera e donna: Scrittura epistolare al femminile tra Quattro e Cinquecento* (Rome: Bulzoni, 1993); *La lettera familiare* (Padua: Liviana, 1985), and *La correspondance: Edition, functions, significations*, I (Aix-en-Provence: Université de Provence, 1984).

7 Angelo De Meis, *Dopo la laurea* (Bologna: Stabilmento Tipografico G. Monti, 1868); Paolo Mantegazza, *Un giorno a Madera* (Milan: Rechiedei 1868); Giovanni Verga, *Storia di una capinera* (Milan: Lampugnani, 1871); Neera (pseud. of Anna Radius Zuccari), *Addio!* (Milan: Brigola, 1877), *Una passione* (1903); Marchesa Colombi (pseud. of Maria Antonietta Torriani), *Prima morire* (1887); Matilde Serao, *Tre donne* (Rome: Voghera, 1905), *Ella non rispose* (Milane: Treves, 1914).

8 I am borrowing the term "cœur-responding" from David Henkin, who noted that, because of a false etymology peddled in Richardson's *Clarissa,* "correspondence" became synonymous with communication of the heart (*The Postal Age,* 99). See also Elizabeth Heckendorn Cook, *Epistolary Bodies: Gender and Genre in the Eighteenth-Century Republic of Letters* (Stanford, CA: Stanford UP, 1996), 16; and Bruce Redford, *The Convers of the Pen: Acts of Intimacy in the Eighteenth-Century Familiar Letter* (Chicago: U of Chicago P, 1986), 1.

9 See Elizabeth C. Goldsmith, *Writing the Female Voice: Essays on Epistolary Literature* (Boston: Northeastern UP, 1989).

10 The above quotation was published in the column "La Conversazione" in *L'Illustrazione Italiana* 27, 5 (7 July 1878): 11.

11 See in particular Zarri, ed., *Per lettera,* and Meredith K. Ray, *Writing Gender in Women's Letter Collections of the Italian Renaissance* (Toronto: U of Toronto P, 2009).

12 For an analysis of letter writing in postmodern culture see Sunka Simon, *Mail-Orders: The Fiction of Letters in Postmodern Culture* (Albany, NY: SUNY Press, 2002).

2. Writing and Reading Letters

1 The use of postal services was naturally a gradual process in the aftermath of Italy's political unification. Partly out of scepticism towards the newly nationalized postal service, and partly out of habit, some people still resorted to private means for the delivery of their mail, as the general director of the Italian Post Office, Giovanni Barbavara, acknowledged in his first annual report: "Ed in vero non tutte le lettere che si affermano perdute vengono impostate da chi le ha scritte, come pure non tutte le lettere si ricevono personalmente dai loro destinatari, ma moltissimi le spediscono e le ricevono per mezzo dei loro commessi e dei loro famigliari" [And really not all the letters which are said to have gone lost were actually mailed by those who wrote them, and also not all letters are personally received, but many send them and receive them by way of an intermediary or family member]: *Prima relazione sul servizio postale in Italia. Anno 1863* (Turin: Tipografia Fodratti, 1864), 15. The sense, however, that one gains from reading the official reports of the Italian Post Office and the burgeoning print media of the time is that there was a prevailing positive perception of the postal service, considered a major sign of progress and modernization. A sign of such progress was the simple fact that many more people than ever before

had at this point in time access to a service until then unavailable to them. Naturally, the question of when Italy's process of modernization started is a complex one, and David Forgacs is indeed correct in defining Italy as a "semi-modern country" until the 1950s. See David Forgacs, *Italian Culture in the Industrial Era, 1880–1980: Cultural Industries, Politics, and the Public* (Manchester: Manchester UP, 1990) and *L'industrializzazione della cultura italiana (1880–2000)* (Bologna: Il Mulino, 1992), 22. My emphasis here is not so much on the actual state of Italy's modernization but rather on how people perceived the postal service, together with the railroad, telegraph, and, later, electricity, as a clear marker of Italy's technological and social progress. On the relation between perception and the actual process of modernization, see also John Davis, "Media, Markets, and Modernity: The Italian Case, 1870–1915," in Ann H. Caesar, Gabriella Romani, and Jennifer Burns, eds, *The Printed Media in Fin-de-Siècle Italy* (Oxford: Legenda, 2011), 10–19.

2 For a presentation of the modern postal service as a revolutionary innovation see Giovanni Paoloni, ed., *Le poste in Italia* (Bari: Laterza, 2005). In the pages of *Nuova Antologia* in 1869, Stefano Jacini also described these changes in terms of revolution: "Sennonchè, per ora, quello che si deve ammettere e puossi desumere dalla Relazione particolareggiata e documentata del 31 gennaio 1867, si è che l'Amministrazione dei lavori pubblici in Italia, nel primo periodo della risurrezione nazionale, malgrado ogni specie di ostacoli e di contrarietà, malgrado la mancanza di preparazione, è stata feconda, come forse nessun'altra d'Europa nell'egual tempo [...] Fra tutti i frutti della rivoluzione italiana, i più utili per la nazione sono quelli che i lavori pubblici hanno procacciato" [Insofar as one can infer from the documented and detailed Report of the 31 of January 1867, the Administration of Public Works in Italy, in the first period of its national resurgence, in spite of all the obstacles and difficulties, in spite of its lack of preparation, has been as productive as in any other European nation during such time]: "Della sistemazione dei Lavori Pubblici in Italia," *Nuova Antologia* 9 (1869): 181. See also John Richard, who in his study of American postal history identifies, in the Post Office Act of 1792, a major event in the shaping of American political thought: *Spreading the News: The American Postal System from Franklin to Morse* (Cambridge, MA: Harvard UP, 1995), 59. Postal innovations were naturally part of a larger technological development of mass communication. For an analysis of how the "revolution" in reading and publishing shaped the circulation of cultural and political ideas in Italy, see Lucy Riall, *Garibaldi: Invention of a Hero* (New Haven, CT: Yale UP, 2007), in particular 128–63.

3 *Edizione nazionale ed europea delle opere di Alessandro Manzoni. Carteggio: Alessandro Manzoni, Claude Fauriel*, ed. Irene Botta (Milan: Centro nazionale studi manzoniani, 2000), 207.
4 In a letter dated 21 February 1821, in ibid., 300.
5 In a letter dated 29 January 1821, in ibid., 283.
6 To give a sense of the variety of tariffs existing throughout the different states: a stamp between 1850 and 1860, when it was adopted by the individual postal systems, cost twenty cents in the Regno di Sardegna, ten in the Granducato di Toscana, and eight and a half cents (two grani) in the Regno delle Due Sicilie. See Quirino Maiorana, "Cenni storici sul graduale sviluppo degli impianti in Italia," *Cinquant'anni di storia italiana I* (Milan: Ulrico Hoepli, 1911); and Enrico Melillo, *Ordinamenti postali e telegrafici degli antichi Stati italiani e del regno d'Italia. Quaderni di storia postale 14* (Prato: Istituto di Studi storici postali, 1991). Because bookstores and publishers (often the same establishment) were located in urban centres, distribution remained a critical aspect of the cultural industry until at least the beginning of the twentieth century. See Forgacs, *L'industrializzazione* 63–4.
7 On 15 December 1860 the Piedmontese government announced a first decree for postal reform, which became effective the following year when the Kingdom of Italy was officially proclaimed on 17 March 1861. In February of 1861, additionally, the new Italian state applied a series of new norms, *Istruzioni speciali provvisorie per il servizio della posta delle lettere*, which also became effective in March of the same year. See Paoloni, ed., *Le poste*, vol. 1: 37.
8 While lacking innovation within its own territory, the Kingdom of the Two Sicilies was very advanced when it came to maritime postal service. On 20 June 1824, the steamboat *Real Ferdinando* left Naples for Palermo, inaugurating the first postal transportation via sea organized and administered by a local – that is, not foreign – entity (the Kingdom of Piedmont established the first transportation of mail by steamboat service in 1835). See Paoloni, ed., *Le poste*, vol. 1: 6–24.
9 The introduction of the fixed tariff, inspired by the successful and revolutionary penny post created by Rowland Hill in England in 1837, helped overcome what was considered a major deterrent to postal communication: the calculation of postal tariff based on weight and distance, which differed from state to state and which often resulted in the mishandling and/or non-delivery of the postal item. The adoption of the stamp gave a real boost to postal activities. Paoloni, ed., *Le poste*, vol. 1: 15.
10 Marina Giannetto, "Ministero delle poste e dei telegrafi (1889–1924)," in Guido Melis, ed., *L'amministrazione centrale. Dall'unità alla repubblica: Le strutture e i dirigenti* (Bologna: Il Mulino, 1992), 157–90.

11 This figure includes also the letters sent in the newly annexed Papal State, but the director, Giovanni Barbavara, clarifies that while the 1871 increment of 9,736,271 accounted also for the letters sent from Rome and the surrounding provinces, the 1872 record of 1,191,087 letters sent in Italy was, instead, more the reflection of sheer added volume of postal delivery. See *Undicesima relazione sul servizio postale in Italia: 1873* (Rome: Tipografia Eredi Botta, 1875), vi. As Barbavara noted in this regard, "E' da notarsi che dal 1870 al 1871 si ebbe un di più di 9,736,271 lettere, ma questo incremento trovava allora ragione, almeno per molta parte, nell'annessione al regno della provincia di Roma, mentre l'aumento verificatosi nel 1872 sul 1871, che fu di 1,191,087 lettere e quello assai più rilevante del 1873 sul 1872 di 4,144,812 lettere, segnano veramente il progressivo svolgersi alla corrispondenza epistolare, e questa, sia che derivi da ragioni di commercio che di industria, sia che tragga la sua origine da relazioni famigliari, è sempre indizio di crescente benessere e di materiale o morale prosperità" [It must be noted that from 1870 to 1871 there were 9,736,271 more letters, but this increment was due, at least for most of it, to the annexation of the province of Rome, while the growth between 1871 and 1872, which amounted to 1,191,087 letters, and even more relevant the one between 1872 and 1873 of 4,144,812 letters, marks the real progress made by epistolary correspondence, which whether it comes from commerce and industry or from personal relationships, is always an indication of a growing well-being as well as of material and moral prosperity]: *Undicesima relazione*, vi.

12 For an analysis of the typologies of nineteenth-century correspondence, see Gino Tellini, *Scrivere lettere: Tipologie epistolari nell'Ottocento italiano* (Rome: Bulzoni, 2002).

13 Carlo Poni, "Contributo al dibattito," *Alfabetismo e cultura scritta nella storia della società italiana* (Perugia: Università degli studi, 1978), 389. Also mentioned in Forgacs, *L'industrializzazione*, 26.

14 See Paolo Macry, "Sulla storia sociale dell'Italia liberale. Per una ricerca sul 'ceto di frontiera,'" *Quaderni storici* 12, 35 (1977): 521–50.

15 For a sample of the letters contained in the CEOD project see the CD-ROM included in Giuseppe Antonelli, Carla Chiummo, and Massimo Palermo, eds, *La cultura epistolare nell'Ottocento* (Rome: Bulzoni, 2004).

16 The transcript and photographic reproduction of these letters may be viewed online at http://ceod.unistrasi.it/integrale.cgi (accessed 28 October 2011).

17 Groundbreaking research in this area of studies remains that of Emilio Franzina, *Merica! Merica! Emigrazione e colonizzazione nelle lettere dei contadini veneti in America Latina 1876–1902* (Milan: Feltrinelli, 1979). For letters of Italian immigrants, see also Antonio Gibelli and Fabio Caffarena,

"Lettere degli emigranti," in Piero Bevilacqua, Andreina De Clementi, and Emilio Franzina, eds, *Storia dell'emigrazione italiana: Partenze* (Rome: Donzelli, 2001), 563–74; and Ilaria Serra, *The Imagined Immigrant: Images of the Italian Emigration to the United States between 1890 and 1924* (Madison, NJ: Fairleigh Dickinson UP), 2009.

18 For an overview of literary representations of emigration in Italian literature see Sebastiano Martelli, "Dal vecchio mondo al sogno Americano. Realtà e immaginario dell'emigrazione nella letteratura italiana," in Bevilacqua, De Clementi, and Franzina, eds, *Storia dell'emigrazione italiana*, 433–87.

19 This as well as subsequent translations in English of Manzoni's *I promessi sposi* are taken from: Alessandro Manzoni, *The Betrothed*, intro. and trans. by Bruce Penman (London: Penguin, 1972), 496.

20 This translation is from Maria Messina, *Behind Closed Doors*, trans. Elise Magistro (New York: The Feminist Press, 2007), 60–1.

21 As Redford wrote, "epistolary discourse accomplishes something even more inventive: it fashions a distinctive world at once internally consistent, vital and self supporting. The letters of a master thereby escape from their origin as reservoirs of facts: coherence replaces correspondence as the primary standard of judgement": *The Convers of the Pen: Acts of Intimacy in the Eighteenth-Century Familiar Letter* (Chicago: U of Chicago P, 1986), 9.

22 Only after 1881 were Italians able to send packages through the Italian Postal Service. See Quirino Majorana, *Cinquanta anni di storia italiana* (Milan: Ulrico Hoepli, 1911), 4.

23 For an overall description of *calendari* and *lunari* see Gianfranco Tortorelli's article "La letteratura popolare," In Ilaria Porciani, ed., *Editori a Firenze nel secondo Ottocento*. Atti del Convegno (13–15 November 1981), (Florence: Leo S. Olschki Editore, 1983), 494–501. See also Gabriella Solari, ed., *Almanacchi, lunari e calendari toscani tra Settecento e Ottocento* (Milan: Giunta regionale toscana/Editrice bibliografica, 1989); Alberto Milano, ed., *Immagini del tempo: 500 anni di lunari e calendari da muro della raccolta Bertarelli* (Bassano del Grappa, Vicenza: Tassotti, 2000); and Giambattista Scirè, ed., *Poste dal cavallo a Internet: Storia illustrata dei servizi postali italiani* (Florence: Giunti Editore, 2008).

24 See Richard Altick, *The English Common Reader: A Social History of the Mass Reading Public 1800–1900* (Chicago: U of Chicago P, 1957); and Silvia Franchini and Simonetta Soldani, eds, *Donne e giornalismo: Percorsi e presenze di una storia di genere* (Milan: FrancoAngeli, 2004), 28–35.

25 Jane Rose formulates a useful distinction between etiquette manuals containing rules on fashion and beauty for mere individual self-advancement

and conduct books, which are, instead, more concerned with the moral and intellectual improvement of society. See her essay "Conduct Books for Women 1830–1860: A Rationale for Women's Conduct and Domestic Role in America," in Catherine Hobbs, et al., eds, *Nineteenth-Century Women Learn to Write* (Charlottesville: UP of Virginia, 1995), 37–58.

26 Although, as Inge Botteri rightly points out, the *istituzioni della sociabilità* changed from the Renaissance to the nineteenth century, as the concepts of society and social relations were drastically transformed, the idea of the social occasion as a main indicator of an individual's integration into an idealized form of social life remained the same. For Botteri's distinction between sociability in antiquity and in the Ottocento, see *Galateo e Galatei: La creanza e l'instituzione della società nella trattatistica italiana tra antico regime e stato liberale* (Rome: Bulzoni Editore, 1999), 15–17.

27 According to Inge Botteri the archetype for this genre, Giovanni della Casa's *Galateo, overo de' costumi*, became outdated after the political unification of the country, when a new consensus on social behaviour and new forms of sociability emerged and superseded older forms. See Botteri, *Galateo e Galatei*; "Lo spazio e il ruolo della famiglia dei galatei dell'Ottocento," in Filippo Mazzonis, ed., *Percorsi e modelli familiari in Italia tra il '700 e'900* (Rome: Bulzoni Editore, 1997), 187–206; and "Le nuove usanze. Società urbana e nuovi costume nei galatei milanesi di fine Ottocento," in Cesare Mozzarelli and Rossana Pavoni, eds, *Milano fin de siècle e il caso Bagatti Valsecchi* (Milan: Guerini e Associati, 1991), 163–80. See also Luisa Tasca, *Galatei. Buone maniere e cultura borghese nell'Italia dell'Ottocento* (Florence: Le Lettere, 2004); Michela DeGiorgio, "Buone maniere in famiglia," in Piero Melograni, ed., *La famiglia italiana dall'Ottocento a oggi* (Rome-Bari: Laterza, 1988), 259–86; and Gabriella Turnaturi, *Signore e signori d'Italia. Una storia delle buone maniere* (Milan: Feltrinelli, 2011).

28 Carol Poster has complained about the paucity of studies on letter writing manuals and suggests that the principal reason for it is one of disciplinarity. Given the tendency of epistolary manuals to straddle different literary genres (from textbooks of rhetoric or grammar to self-help books), scholars specialized in specific fields tend to marginalize this type of text. See "Introduction," in Carol Poster and Linda Mitchell, eds, *Letter Writing Manuals and Instruction from Antiquity to the Present: Historical and Bibliographic Studies* (Columbia: U of South Carolina P, 2007), 2.

29 Conduct books became part of that vast production of didactic material written by women whose "natural" predisposition towards teaching was considered a crucial component in the process of education of the masses. Gabriella Turnaturi suggests that these women writers created a school

of good manners, meaning that "è da queste donne che la borghesia va a scuola di buone maniere" [it is from these women that the bourgeoisie learns the basic precepts of good manners] (*Signore e signori d'Italia*, 44). In publishing as well as in schools, women flourished under the assumption that they were natural teachers.

30 On the relationship between Cesare Causa and Adriano Salani, see Emilio Faccioli, "Un editore popolare di orientamento Adriano Salani," in Porciani, ed., *Editori a Firenze nel secondo Ottocento*, 367–80.

31 See Adriano Salani in his preface "Al lettore" in *Il segretario italiano osia modo di scrivere lettere sopra ogni sorta di argomenti per affari di commercio amici e conoscenti per feste ed anniversari di famiglia lettere di raccomandazione, di domanda ecc. Con l'aggiunta: Di modelli per cambiali, suppliche e ricevute, ricorsi al re, ai ministri ed altri funzionari; e la spiegazione grammaticale per scrivere ben corretto e con buona ortografia* (Florence: Adriano Salani, 1903).

32 The same judgment is made by a bookseller in Milan, Mr Bruciati who, talking about his buyers' reading tastes, says, "Una classe speciale che ha gusti particolari è quella delle serve che leggono solo il *Re dei cuochi*, il *Segretario galante*, l'*Oracolo delle donne*, *Genoveffa* dello Schmid, *Elisabetta in Siberia* di Cottin" [A special class of readers is that one of maids who read only "The King of Chefs," "The Gallant Secretary," "Women's Oracle," Schmid's "Genoveffa," and "Elizabeth in Siberia" by Cottin]: *I libri più letti dal popolo italiano: Primi risultati dell'inchiesta promossa dalla Società Bibliografica Italiana* (Milan: Società Bibliografica Italiana, 1906), 11.

33 As one of the editors admitted, "La presente operetta pertanto non è che il Piccolo segretario galante, emendato in alcuni luoghi, considerevolmente aumentato e con diligenza corretto" [This volume is simply the *Piccolo segretario galante* amended in several parts, considerably augmented and diligently corrected] *Piccolo segretario galante o raccolta di lettere e biglietti amorosi aggiuntavi la corrispondenza di due infelici amanti* (Milan: Tip. Flli Bietti e G. Minaccia, 1882), 3.

34 See Faccioli, "Un editore popolare," 374–6.

35 A reprint of the Salani catalogue maybe found in Ada Gigli Marchetti, *I libri buoni e a buon prezzo. Le edizini Salani (1862–1986)* (Milan: Franco Angeli, 2011).

36 During the Risorgimento, patriots considered newspapers an essential vehicle for the circulation of ideas and development of a national consciousness. In 1848, for instance, after the riots of the *Cinque giornate* in Milan, a local newspaper stated with the fervour of patriotic eloquence that, "Il giornalismo è il sole che dirada le nebbie dell'ignoranza, che svolge e matura i grandi sistemi di civilizzazione; è la luce che scopre e addita i

bisogni della società [...] il faro che guida pel vasto oceano della politica, dell'economia pubblica, della scienza, dell'arte" [Journalism is the sun that dissipates the fog of ignorance, that develops and matures the great systems of civilization; it is the light that unveils and shows the needs of society ... the lighthouse that guides through the vast ocean of politics, public economy, science and art]: Paolo Murialdi, *Storia del giornalismo italiano* (Bologna: Il Mulino, 2006), 55–6.

37 On Italian salon life, see: Maria Luisa Betri and Elena Brambilla, eds, *Salotti e ruolo femminile in Italia tra fine Settecento e primo Novecento* (Venice: Marsilio, 2004); Maria Teresa Mori, *Salotti: La sociabilità delle elite nell'Italia dell'Ottocento* (Rome: Carocci, 2000); and Maria Iolanda Palazzolo, *I salotti di cultura nell'Italia dell'Ottocento. Scene e modelli* (Milan: FrancoAngeli, 1985). For a comparative study of French and Italian salons, see Elena Brambilla, "Donne, salotti e Lumi: Dalla Francia all'Italia," in Andreina De Clementi, ed., *Il genere dell'Europa. Le radici comuni della cultura europea e l'identità di genere* (Rome: Biblink Editori, 2003), 57–92; and Benedetta Craveri, "Salons francesi e salotti italiani: Proposte di confronto," in Betri and Brambilla, eds, *Salotti e ruolo femminile*, 539–44.

38 On the specific traits of the the nineteenth-century Italian salon, see Mori, *Salotti*, in particular, "Le regole del gioco,"27–59.

39 *Corriere della Sera*, for instance, in the 1870s ran several letter-columns, including "Lettera aperta alle signore" and "Lettera aperta ai bambini," both written by Marchesa Colombi; and Edmondo De Amicis wrote his articles as a foreign correspondent from Spain in 1872 (for *La Nazione*) and from Paris in 1878 (for *L'Illustrazione Italiana*) also in letter form. In general, correspondence from other cities was often presented as letters.

40 On *La Toelette* see Franchini and Soldani, eds, *Donne e giornalismo*, 28, 184, 311–16; and Roberta Turchi, "La Toelette," in Silvia Franchini, Monica Pacini, and Simonetta Soldani, eds, *Giornali di donne in Toscana: Un catalogo, molte storie (1770–1945)*, vol. 1 (Florence: Leo S. Olschki, 2007), 97–101.

41 "Lettera alla Gentile Signora Elisabetta Caminer," *La Toelette* 7 (1771): ix.

42 Ruth Messbarger defines the *Settecento* as "The Century of Women," borrowing this expression from Pietro Chiari. See her *The Century of Women: Representations of Women in Eighteenth-Century Italian Public Discourse* (Toronto: U of Toronto P, 2002), 3.

43 *La Toelette* 8 (1771): i.

44 "Lettera alla Signora Elisabetta Caminer," *La Toelette* 12 (1771): vi.

45 Ibid., iv.

46 According to a survey conducted in 1811, *Corriere delle Dame* sold around 700 copies. See Franchini and Soldani, eds, *Donne e giornalismo*, 88;

Murialdi, *Storia del giornalismo italiano,* 31–2, 54. In the period spanning the decade 1850–60, *Corriere delle Dame* reached 1,000 copies, while its sister publication, *La Ricamatrice,* sold about 2,000 copies, a notable number if one thinks that a newspaper like *Il Crepuscolo* sold no more than 2,600 copies. See Patrizia Landi, "Non solo moda. Le riviste femminili a Milano (1850–1859)," in Nicole Del Corno and Alessandra Porati, eds, *Il giornalismo lombardo nel decennio di preparazione all'unità* (Milan: FrancoAngeli, 2005), 223; and Giuseppe Farinelli, Ermanno Paccagini, Giovanni Santambrogio, and Angela Ida Villa, eds, *Storia del giornalismo: Dalle origini ad oggi* (Turin: UTET, 2004), 169.

47 Giuditta Lampugnani, "Corrispondenze," *La Ricamatrice* 7, 2 (16 January 1855), 20.

48 Giuditta Lampugnani begins the rubric "Corrispondenze" of 16 January 1855 with a reproach to her readers who signed their letters only with their initials, in fear, perhaps, of compromising their names and that of their families. She wrote, "Abbiamo un rimprovero per alcune delle gentili nostre corrispondenti. Primeriamente perché le loro lettere non portano sottoscrizione, poi perché, non contente di nascondersi sotto iniziali, che per noi equivalgono ad un anonimo assoluto, anziché scrivere per proprio conto, si dicono interpreti delle loro amiche, o dei loro consiglieri e segretari intimi, per avvalorare quelle osservazioni di cui forse nel fondo della loro coscienza riconoscono l'illegalità e l'ingiustizia [...] Proponendo la corrispondenza col giornale alle giovani come mezzo d'esercizio epistolare e di giudiziose riflessioni, raccomanderemo la spontaneità e la freschezza che deriva dall'intimo convincimento di ciò che si scrive" [We reproach our dear correspondents, mainly because their letters are not signed, as if it were not enough to hide behind initials, which for us signify absolute anonymity; instead of writing as themselves, they say they are speaking of their friends, or of their counselling and intimate secrets, in order to validate those observations which they recognize in conscience as being illegal and unjust ... By proposing this correspondence with our newspaper to young women as a means of epistolary exercise and wise observations, we would recommend the spontaneity and freshness which derive from the intimate conviction of what is written]: *La Ricamatrice* 8, 2 (16 January 1855), 20.

49 For an analysis of the role of the family in relation to the development of a national rhetoric in the nineteenth century see Ilaria Porciani, "Famiglia e nazione nel lungo Ottocento," in Ilaria Porciani, ed., *Famiglia e nazione nel lungo Ottocento italiano: Modelli, strategie, reti di relazioni* (Rome: Viella, 2006), 15–53.

50 Giuditta Lampugnani, *La Ricamatrice*.8, 2 (16 January 1855).
51 Giuditta Lampugnani, *Giornale delle Famiglie: La Ricamatrice* 16, 20 (1 August 1862).
52 "Dello stile epistolare I," *La Ricamatrice* 7, 5 (1 August 1854), 133–4.
53 "Bibliografia: Guida allo studio delle belle lettere ed al comporre con una manuale dello stile epistolare di Giuseppe Picci. Milano per G. Gnocchi, 1855," *La Ricamatrice* 8, 21 (1 November 1855), 202–4.
54 *La Ricamatrice* 5, 8 (16 April 1852), 60.
55 Patrizia Zambon identified Nievo under this pseudonym on the basis of annotations found among the author's papers on the earnings he received for his writings. See Ippolito Nievo, *Scritti giornalistici alle lettrici*, ed. Patrizia Zambon (Lanciano, Chieti: Casa Editrice Rocco Carabba, 2008), 65.
56 *La Ricamatrice* 12, 2 (16 March 1859), 47. Nievo here was adding to an ongoing debate in *La Ricamatrice* on the linguistic aspect of women's education. A couple of months earlier, the periodical had published an epistolary exchange titled "I misteri del linguaggio. Lettere ad Emilia," in which the author, "Amelia," defended the importance of standard Italian in the cultural formation of young women. For the transcript of Nievo's article and information about it see Nievo, *Scritti giornalistici alle lettrici*, 279–82 and 417–20.
57 *La Ricamatrice* 10, 4 (16 February 1847), 31.
58 *La Ricamatrice*10, 6 (16 March 1857), 50.
59 Ibid.
60 To give a sense of this competition, around the same time that *Passatempo/ Giornale delle Donne* was published, these other newspapers were after the same small female readership: *La Donna* (1868–81), *Margherita* (1878–1921), *Giornale delle Signore Italiane* (1878–96), *La Missione della Donna* (1872–95), *La Moda. Giornale delle Dame* (1878–91), *Cornelia* (1872–88), *Cordelia* (1881–1943), and *L'Aurora* (1872–?). See Carla Dappio, "I Periodici femminili dell'800 in due biblioteche romane," *Memoria* 5 (1982):118–21; Franchini and Soldani, eds, *Donne e giornalismo*; and Annarita Buttafuoco, "Educazione e modelli di emancipazione nella stampa politica femminile del secondo Ottocento," *Cronache femminili.Temi e momenti della stampa emancipazionista dall'Unità al fascismo* (Siena: Dipartimento di studi storico-sociali, 1988), 21–52.
61 *Passatempo: Letture per il gentil sesso* 2 (February 1869), Appendix 1.
62 These letter-articles were published again in 1876, always in *Giornale delle Donne*, a sign that the interest on this topic was still strong.
63 Neera, "La donna libera: Idee di Neera." *L'Illustrazione Italiana* III, 23 (2 April 1876), 367.

64 Marchesa Colombi, "La donna povera. Lettera della Marchesa Colombi alla signora Neera," *L'Illustrazione Italiana* III, 25 (16 April 1876), 398.
65 Matilde Serao, "La politica del cuore," *Corriere del Mattino* (30 April 1880). This article was in response to an open letter written to her by the director of the same newspaper, Martino Cafiero, published on 27 April.
66 Anna Maria Mozzoni, "Lettera aperta di Anna Maria Mozzoni a Matilde Serao," *La Lega della Democrazia* I, 132 (15 May 1880).
67 See the transcript of both articles in the Appendix: Matilde Serao, "La serva," *La Stampa* XXXIX, 56 (25 February 1905), 1–2; and Marchesa Colombi, "La padrona," *La Stampa* XXXIX, 60 (1 March 1905), 1–2.

3. Fictionalizing the Letter

1 Quoted in an article by Angiolo Orvieto, "Come è nato e che cosa è il nuovo romanzo di Neera." *L'Illustrazione Italiana* XXXIV, 20 (19 May 1907): 404.
2 In a letter sent on 5 August 1873 to Giovanni Verga, Emilio Treves announced that the publication of *Eva* and *Storia di una capinera* had been received with enthusiasm in Milan: "Vi piace l'edizione? Qui i due romanzi hanno suscitato un vero entusiasmo; al vostro ritorno, le signore vi ruberanno. Nei circoli letterari, non si parla d'altro" [Do you like the edition? Here your two novels have sparked great enthusiasm; women will snatch you when you return. In literary circles, this is all they speak about] (Raya 28).
3 An exception to this general approach is Irene Gambacorti's excellent analysis of Verga's epistolary novel, *Verga a Firenze: Nel laboratorio di Storia di una capinera* (Florence: Le Lettere, 1994), although other recent critical assessments of this novel still consider it part of a failed initial search for an authorial artistic voice, relegating it to what has been traditionally interpreted as Verga minore. See, for instance, Simona Cigliana, *L'immaginario di Verga* (Rome: Salerno, 2006).
4 A starting point in the critical neglect of Verga's early literary production may be identified with Luigi Russo's publication of 1919 on Verga's oeuvre. Russo, lamenting the lack of an appropriate "riconoscimento storico" [historical recognition] of Verga's work on the part of critics and recognizing the importance of criticism, "la seconda vita di un'opera, è il suo futuro, e la sua perpetua rivelazione" [its second life, its future, and its perpetual revelation], endorsed Verga's *verista* production, because it eluded biographical connections and the reader's participation, "non lasciava margine ai capricci del lettore e alla sua abilità di psicologista e fantasista" [it did not offer any scope for the reader's caprices and psychological and

fantasizing abilities]. See Luigi Russo, *Giovanni Verga* (Bari: Laterza, 1959), 22, 6.

5 For an analysis of the relationship between author and public in post-unification Italy, see Alberto Cadioli, "Autori e pubblico," in Franco Brioschi and Costanzo di Girolamo, eds, *Manuale di letteratura italiana: Storia per generi e problemi,* vol. IV (Turin: Bollati Boringhieri, 1993–6), 116–40, at 123.

6 In this regard, Scarfoglio wrote, "E poi la libertà, non dico della politica, ma dell'arte, è distrutta. Siamo anche in arte, sotto l'imperio della maggioranza; e la maggioranza vile che non sa nulla, che non intende nulla, che non desidera se non cose sciocche e volgari, vuole anch'essa i suoi istrioni, i suoi glorificatori, i solleticatori de' suoi istinti o cattivi o malsani o imbecilli; o se non è contenta degli istrioni, non li paga" (206–7) [And then, the liberty, I wouldn't say political liberty, but that for the arts, is destroyed. We are also, in the arts, under the rule of the majority; the vile majority, that knows nothing, that understands nothing, that desires nothing if they be not foolish or vulgar things, that wants its actors, its glorifiers, those who tickle its instincts, whether bad or unwholesome or idiotic; or if it is not pleased by the actors, it does not pay them].

7 In *Dell'origine e dell'ufficio della letteratura,* a lecture delivered on 22 January 1809 to professors and students of the university of Pavia, Ugo Foscolo had already raised similar concerns, but both his lecture and its later publication had only a limited audience of students and literati, while Bonghi's articles, published in a newspaper, enjoyed a wider readership.

8 Matilde Serao, for instance, towards the end of the nineteenth century defined the "Italian" novel as a proposition for the future and not the present, for what writers wrote then were regional renditions of a possible national novel. She said, "Il romanzo italiano non può esistere per ora; tutti i romanzi che noi facciamo sono parti, elementi, coefficienti del futuro romanzo italiano integro e perfetto; essi sono, se non per altro, per l'argomento essenzialmente regionali" [The Italian novel cannot for now exist; all the novels we write are part, elements, coefficients of the perfect, integral Italian novel of the future; they are, if anything else, essentially regional in light of their themes]. See Ugo Ojetti, "Matilde Serao," in *Alla scoperta dei letterati,* 234.

9 See Gianfranco Tortorelli, "I libri più letti dal popolo italiano: Un'inchiesta del 1906," *Studi di storia dell'editoria italiana* (Bologna: Patron editore, 1989): 153–69; and *I libri più letti dal popolo italiano: Primi risultati della inchiesta promossa dalla società bibliografica italiana* (Milan: Società Bibliografica Italiana, 1906).

10 *Storia di una capinera* was also presented in 1874 as a gift to subscribers of another of Treves's periodicals, *Museo di Famiglia*. See Gambacorti, *Verga a Firenze*, 131–2.

11 *Storia di una capinera* appeared in *Corriere delle Dame. Giornale di Moda ed Amena Letteratura* and not in *La Ricamatrice* as Federico De Roberto erroneously reports in *Casa Verga*. See Gianluigi Berardi, "La capinera ritrovata," in Giovanni Verga, *Storia di una capinera* (Pordenone: Edizioni Studio Tesi, 1985), ix–xiii. On the general editorial history of the novel, see: Gambacorti, *Verga a Firenze*, 127–34; Gino Tellini, "Il monologo della 'povera capinera,'" *L'invenzione della realtà. Studi verghiani* (Pisa: Nistri-Lischi, 1993), 172–90; Remo Cacciatori, "Il successo della 'Storia di una capinera' e la narrativa patetica ottocentesca," *Scrittore e lettore nella società di massa. Sociologia della letteratura e ricezione lo stato degli studi* (Trieste: Edizioni Lint, 1991), 313–42; Andrew Wilkin, "Giovanni Verga's 'Storia di una capinera' – 100 Years On," *Modern Languages* 52 (1971): 177–81; Federico De Roberto, Storia della 'Storia di una capinera,' *Casa Verga e altri saggi verghiani*, ed. Carmelo Musumarra (Florence: Le Monnier, 1964), 163–79.

12 In those same years, Verga wrote *Eva*, which came out later with Treves in 1873 and which the author presented in the introduction with words that express similar feelings of denunciation: "La civiltà è il benessere, e in fondo ad esso, quand'è esclusivo come oggi, non vi troverete altro [...] Viviamo in un'atmosfera di Banche, e di Imprese Industriali, e la febbre di piaceri è la esuberanza di tal vita. Non accusate l'arte, che ha il solo torto di aver più cuore di voi, e di piangere per voi i dolori dei vostri piaceri. Non predicate la moralità [...] voi che fate scricchiolare allegramente i vostri stivali inverniciati dove folleggiano ebbrezze amare, o gemono dolori sconosciuti, che l'arte raccoglie e che vi getta in faccia" [Civilization means wealth, and at its core, when it is so exclusive as it is today, you will not find anything else. We live in an environement of banks, industrial companies, and the desire for pleasure is the impetus of such life. Don't blame art, which has the only fault of having a bigger heart than you all have, and of lamenting the pain caused by your pleasures. Don't preach morality ... you who gaily walk with your squeaking patent-leather shoes where bitter intoxications frolic or unknown pains moan, which art gathers and throws in your face]. Cited in Isabella Gherarducci and Enrico Ghidetti, eds, *Guida alla lettura di Verga* (Florence: La Nuova Italia, 1994), 80.

13 I would like to thank Margherita Verdirame for helping me gain access to the reviews of *Storia di una capinera* written in the early 1870s.

14 Quoted from an article by Giorgio Baseggio that appeared in "Rassegna bigliografica" in *Perseveranza* (14 September 1873).

15 See reviews by C. D'Ormerville in *Il Pungolo* (7 August 1873) and F. D'Arcais in *L'Opinione* (15 August 1873).
16 See Alfonso Scirocco, "Il dibattito sulla soppressione delle corporazioni religiose nel 1864 e i 'Misteri del chiostro napoletano' di Enrichetta Caracciolo," *Clio: Rivista trimestrale di studi storici* 2, 2 (1992): 215–33.
17 In the same periodical, Percoto published a few years earlier (1857–8) an epistolary novel titled *Il giornale di mia zia*, an educational narrative infused with historical references to the Risorgimento that has been interpreted by critics as one of Percoto's attempts to move from the format of the short story to that of the novel. See Rossana Caira Lumetti's introduction to Caterina Percoto, *Il giornale di mia zia* (Rome: Bulzoni, 1984), 7–42.
18 Caterina Percoto, *Giornale delle Famiglie* 7bis (1864), 77. The full transcript of Caterina Percoto's "Memorie di convento" is available at the end of this book in the Appendix.
19 Percoto, "Memorie di convento: Lettera terza," *Giornale delle Famiglie* 10bis (16 May 1864), 112.
20 Percoto, "Memorie di convento: Lettera terza," 113.
21 Percoto, "Memorie di convento: Lettera quarta," *Giornale delle Famiglie* 12bis (16 June, 1864), 138.
22 On the editorial history of Percoto's "Memorie di convento," see Rossana Caira Lumetti, *Le umili operaie: Lettere di Luigia Codemo e Caterina Percoto* (Naples: Loffredo, 1995), 34–40, at 40.
23 Francesca Billiani offers an interesting reading of Tarchetti's trilogy on love as an alternative "gothic" expression of the prevailing literary model offered by the historical novel. To the romantic and providential love story of Renzo and Lucia portrayed in Manzoni's *I promessi sposi*, Tarchetti opposed the "depiction of delusional artists and their ambivalent sexual desire towards women." See Billiani's article "Delusional Identities" (in particular, 494–5).
24 Luigi Capuana, remembering the years spent in Florence, wrote in a letter to Salvatore Farina in 1884 that "Mi ero buttato nella filosofia, mi pasceva di Hegel e di positivismo […] O le giornate e le nottate trascorse sopra un libro del De Meis […] quando quel libro – *Dopo la laurea* – giungeva proprio in tempo per trascinarmi più accosto alla realtà e darmi un equo senso della vita!" [I threw myself into philosophy, I nurtured myself with Hegel … and positivism. O the days and nights spent reading De Meis's book … when that book – *Dopo la laurea* – arrived just in time to sweep me closer to reality and give me a just sense of life!]. Quoted in Bigazzi, 102.

25 In a letter to Capuana of 5 April 1873, Verga wrote, "Sì, Milano è proprio bella, amico mio, e credimi che qualche volta c'è proprio bisogno di una tenace volontà per resistere alle sue seduzioni, e restare al lavoro. Ma queste seduzioni istesse sono fomite, eccitamento continuo al lavoro, sono l'aria respirabile perché viva la mente; ed il cuore, lungi dal farci torto non serve spesso che a rinvigorirla. Provasi davvero la *febbre di fare*; in mezzo a codesta folla briosa, seducente, bella, che ti si aggira attorno provi il bisogno di isolarti, assai meglio di come se tu passi in una solitaria campagna. E la solitudine ti è popolata da tutte le larve affascinanti che ti hanno sorriso per le vie e che sono diventate patrimonio della tua mente" [Yes, Milan is really beautiful, my friend, and believe me sometimes there is a real need to be strong to resist its seductions and be able to work. But these seductions are themselves a constant excitement to work, they are the air we breathe because the mind is alive, and the heart, far from doing any wrong, fosters often its reinvigoration. One truly feels the *fever of doing*; in the middle of such a spirited, seductive, and beautiful crowd which revolves around you, you feel the need to isolate yourself, even more than if you were in a solitary countryside. And your solitude is populated by all the fascinating larvae which smiled at you in the streets and that have now become the patrimony of your mind] (Raya, *Carteggio* 25–6). And in another letter, also to Capuana (dated 13 March 1874), Verga described the impact that a city like Milan could have on a man arriving in the big city from Sicily: "Io immagino te, venuto improvvisamente dalla quiete tranquilla della nostra Sicilia, te artista, poeta, matto, impressionabile, nervoso come me, a sentirti penetrare da tutta questa febbre violenta di vita in tutte le sue ardenti manifestazioni, l'amore, l'arte, la soddisfazione del cuore, le misteriose ebbrezze del lavoro, pioverti da tutte le parti, dall'attività degli altri, dalla pubblicità qualche volta clamorosa, pettegola, irosa, dagli occhi delle belle donne, dai facili amori, o dalle attrattive pudiche" [I imagine you, suddenly arrived from the quiet calm of our Sicily, you artist, poet, madman, impressionable and nervous like me, and feeling overwhelmed by all this fever of work, falling on you from all sides, from the activities of others, from the publicity at times noisy, gossipy and angry, from the eyes of beautiful women, from easy loves, and timid attractions] (Raya, *Carteggio* 30).

26 In a letter written from Florence on 4 May 1869, Verga wrote, "Se mi spedite qualche giornale colla parola scritta dietro la fascetta, metteteci solo la data e mai la firma, poiché nel caso che si scoprisse io rifiuterò il giornale e non trovandosi chi lo dirige non potremo pagare la multa, così voi pure in ogni caso lo respingerete" [If you send me newspapers with words written behind the paper strip, write only the date and never your

name, so that in case it should be discovered I will refuse to take the newspaper and if they do not know who the sender is we will not pay the fine, and in any case you will also refuse it]. In Verga, *Lettere sparse*, 9.

27 On 12 July 1872, Verga wrote from Catania to Capuana, "Mi rallegro con te della tua operosità e avrei desiderato venire a stringerti la mano e fare con te una di quelle belle chiacchierate d'arte che sarà pure sonnabulismo ma è pure il bel sogno" [I congratulate you for your productivity and I would have liked to be able to come, shake your hand, and have one of those beautiful art conversations which might very well be somnambulism but is also a beautiful dream] (Raya, *Carteggio* 20).

4. Cœur-responding with Her Readers

1 Niccolò Tommaseo, *La donna: Scritti vari editi e inediti* (Milan: Ditta tipografica libraria Giacomo Agnelli, 1868), 296.

2 Matilde Serao, "Come un piombo," *La vita è così lunga* (Milan: Fratelli Treves, 1918), 20–1.

3 The arbitrariness of such division is reflected in the lack of a general consensus on when the real "downfall" of Serao's creativity began. For Tommaso Scappaticci, *Il paese di cuccagna* (1890) constitutes the dividing work between the two phases (*Introduzione a Serao* [Rome-Bari: Laterza, 1995], 103), while for Pancrazi, 1905 is the date marking the beginning of her narrative decline. See Pietro Pancrazi, ed., *Serao* (Milan: Garzanti, 1944), ix.

4 In a letter written to Count Primoli, Serao said about Bourget, "Apprezzo queste finissime analisi delle anime inferme, io ammiro assai il Bourget [...] quanto è bello quel mondo di creature amorose e scettiche, di esseri elevati quasi sublimi. Quello che più mi piace in questo volume è l'ultimo pezzo di *profils perdus*, quella donna sola, altamente drammatica, che singhiozza nella notte, nella rovina della sua esistenza" [I appreciate the fine analyses of these invalid souls. I greatly admire Bourget ... that world of amorous and sceptical creatures, of elevated almost sublime beings, is so beautiful. What I love the most in this volume is the piece of *profils perdus*, that woman alone, crying in the night and in her ruinous existence, is so highly dramatic]. Quoted in the introduction to M. Serao, *La conquista di Roma*, ed. Wanda De Nunzio Schilardi (Rome: Bulzoni, 1997), xxi.

5 As Suzanne Stewart-Steinberg, for instance, wrote, "Anxiety does in fact describe the post-1860 moment [...] This anxiety is at once political and cultural and, within the context of my argument, fundamental to Italian modernity rather than an impediment to it": *The Pinocchio Effect*, 2.

6 See Serao's interview in Ugo Ojetti, *Alla scoperta dei letterati*, 237.

7 For Cantù, creating a new nation meant essentially creating a new consciousness, which he called "coscienza del dovere" [consciousness of our duty], and which "non meno delle altre facoltà, dev'essere educata. A ciò servono le istruzioni che ci danno i genitori sin dall'infanzia, le prediche, i discorsi e gli esempi de' migliori, le prudenti letture, lo studio della morale, che è appunto la scienza la quale addita il bene e il modo di dirigere le libere azioni verso il fine dell'uomo" [no less than other faculties, must be educated. To this end are aimed the instructions given to us by our parents since childhood, the preaching, the speeches, and examples given by the best people, prudent readings, the study of morality, which is actually the science that shows us the good and the way to make our free actions the goal of human life]: *Buon senso e buon cuore* (Milan: Tipografia editrice Giacomo Agnelli, 1872), 4.

8 On 27 April 1880, Martino Cafiero wrote an open letter to Matilde Serao in *Il Corriere del Mattino*, to which Serao replied with her "La politica del cuore. Al signor Cafiero": *Il Corriere del Mattino* (30 April 1880), 3. Only a few days before, Serao had, in the pages of the daily *La Lega della Democrazia*, engaged in a similar exchange of letters, this time with Agostino Bertani, in which she stated her belief in what she called women's "politics of the heart." To Bertani's accusation that Serao had written a short story, "Piccola anima" [Small soul], focused on the prince of Naples and intended, according to Bertani, to gain favour with the royal family, Serao replied with a defence of her artistic freedom, concluding, "Ma è anche assurdo quello a cui mi ha costretta l'onorevole Bertani. Come scrittore ho dovuto spiegargli i miei intendimenti artistici; speciale tormento, perché ciò che crea la fantasia non si sminuzza, perché l'arte si comprende ma non si spiega. Come donna, poi, mi costringe a dirgli che in fatto di politica le donne non hanno una opinione, hanno un sentimento. Quindi amano la monarchia. È la politica del cuore – ed è anche la mia" [And it is absurd to see what the deputy Bertani has forced me to do. As a writer, I had to explain my artistic intentions; a special torment, because what is created by fantasy cannot be dissected, because one must understand and not explain art. As a woman, then, he forced me to say that in political matters women do not have an opinion, they have a sentiment. Therefore they love the monarchy. It's the politics of the heart, and it is mine as well]. Matilde Serao, "Lettera della signorina Serao," *La Lega della Democrazia* I, 114 (27 April 1880). This letter was a reply to Agostino Bertani's article "Un traviamento letterario," in *La Lega della Democrazia* I, 109 (22 April 1880). This exchange ended with a letter by Bertani titled, "L'on. Bertani alla signorina Serao," also published in *La Lega della Democrazia*, I, 115 (28 April 1880), in which the republican and

radical member of Parliament invited Serao to keep her artistic focus on common people rather than on members of the nobility.

9 On Serao's antifeminism, see, for instance, Wanda De Nunzio Schilardi, "L'antifemminismo di Matilde Serao," *L'invenzione del reale. Studi su Matilde Serao* (Bari: Edizioni Palomar, 2004), 59–94.

10 "Ormai a furia di essere reali, a furia di scartare ogni idea poetica, si diventa aridi e noiosi e lo stesso bel sesso si mette a far da *positivo*; onde la società moderna pare cammini sotto un cielo di ferro, in un ambiente di elettrico e di vapore, con un orizzonte di carta-moneta. Quindi, per reazione, si nota un trasporto verso quello che rappresenta ancora la poesia, i fiori, e l'amore" [These days by being too real, by discounting every poetic idea, one becomes arid and boring and even the genteel sex begins to be positivistic; and modern society seems to walk under an iron sky, in an electric steam environment, with a horizon made of paper money. Therefore, for reaction, one begins to feel carried towards what still constitutes poetry, flowers, and love]: *Giornale di Napoli* (21 December 1877); quoted in Scappaticci, *Introduzione a Serao*, 12.

11 See Matilde Serao, *Sognando* (Catania: Giannotta, 1906), 23, 37.

12 For my use of the term "cœur-responding," see note 8 in chapter 1.

13 See Janet Altman, who writes, "The letter novel is one of the first genres constituted by discovery of a medium and exploration of its potential": *Epistolarity: Approaches to a Form*, 211.

14 On the notion of fantasy as the projection of a sense of wholeness see Joan Wallach Scott, *The Fantasy of Feminist History* (Durham: Duke UP, 2011).

15 Matilde Serao,"Lettera d'Amore," *Novelle sentimentali* (Livorno: S. Belforte E C., Editori, 1902), 13–14.

16 Serao defended the Italian woman against accusations of decadence. She wrote, "È convenzionale il sempiterno paragone fra la donna italiana e la russa, la tedesca, l'inglese, paragone con relativo piagnisteo sulla decadenza della donna italiana: mentre ognuno sa e non vuole confessare che la italiana è molto più seria, molto più onesta, molto più donna che le anglosassoni o le slave" [It's the usual eternal comparison between the Italian woman and the Russian, German and English woman. A comparison accompanied by the usual whining about the decadence of the Italian woman, while everybody knows but doesn't want to confess that the Italian is much more serious, honest, much more woman than the English or the Slavic ones]: quoted in Pier Carlo Masini, *Eresie dell'Ottocento. Alle sorgenti laiche, umaniste e libertarie della democrazia italiana* (Milan: Editoriale Nuova S.p.A., 1978), 277. On the development of an international negative reputation of Italian women, especially in connection to the practice

of *cicibeismo*, see Silvana Patriarca, *Italian Vices. Nation and Character from the Risorgimento to the Republic* (Cambridge: Cambridge UP, 2010), while on the allegorical representation of the Italian nation as a woman see Alberto Banti, *La nazione del Risorgimento* (Turin: Einaudi, 2000).

17 Serao concludes the story with the following: "Quanti anni sono trascorsi? Non so; non mi ricordo più. Molti certamente. Ma per quanto tempo sia passato, per quanto io abbia studiato, non mi sono mai potuta accertare in quale delle tre lettere quella donna mentiva" [How many years have passed? I don't know, I can't remember anymore. Many for sure. Regardless of how much time has passed, as much as I have studied, I could never determine in which of the three letters the woman was lying]: in *Fior di passione* (Milan: Baldini & Castoldi, 1899), 204.

18 Serao, *Lettere d'amore* (Il perché della morte), 3–4.

19 Already in the 1880s Serao understood the state of misery in which these human billboards not only worked but also lived. A few decades later, in the 1930s the so-called sandwichmen were normally recruited from the ranks of clochards. See Susan Buck-Morss, "The Flaneur, the Sandwichman and the Whore."

20 The article, signed with the pseudonym "Chiquita," appeared on 1 December 1886 in *Corriere di Roma,* under the column "Per le signore."

21 Serao, "Falso in scrittura," *Fior di passione*, 197–9.

22 Serao, "L'ultima lettera," *Gli amanti*, 283–4.

23 Serao, *Lettere d'amore*, 55–6.

24 Serao, *Ricordando Neera* (Milan: Treves, 1920), 19–20. Also quoted in Elena Candela, *Amor di Parthenope* (Pisa: Edizioni ETS, 2008), 117–18.

25 Alberto Mario Banti famously defined this bond as a familial network: "relazione parentale."

26 Federigo Verdinois, *Profili letterari* (Florence: Le Monnier, 1949), 180, and quoted in Pupino, *Notizie del reame*, 42.

27 Appeared in the rubric "Per le signore" published in *Corriere di Roma* on 29 March 1887. The article was signed by Serao with the pseudonym "Chiquita."

28 See also Elena Candela, "Matilde Serao: 'A furia d'urti, di gomitate' verso la modernità," *Matilde Serao: Le opere e i giorni* (Naples: Liguori, 2006), 55–77, where she analyses Serao's relationships with other contemporary authors and her contribution to the modernization of Italian literature.

29 For an analysis of the representation of female readership in nineteenth-century Italy, see Maria Iolanda Palazzolo, "Le donne e la lettura," *Domensioni e Problemi della Ricerca Storica* 2 (1991): 87–96; and Ann H. Caesar,

"Women Readers and the Novel in Nineteenth-Century Italy," *Italian Studies* 56 (2001): 80–97.

30 *Prose, e poesie del Signor Abate Antonio Conti, patrizio veneto* (Venice: Pasquali, 1756), lxi; also quoted in Luciano Guerci, *La discussione sulla donna nell'Italia del Settecento: Aspetti e problemi* (Turin: Tirrenia Stampatori, 1987), 144.

31 The figure of the "mother" is central, for instance, in Risorgimento female iconography. Adelaide Cairoli and Eleonora Ruffini became the symbol of the sacrifice and martyrdom suffered by Italian patriots.

32 In *Enciclopedia italiana e dizionario di conversazione. Opera originale*, vol. 7 (Venice: Tasso, 1844), 576; also quoted in Porciani, ed., *Le donne a scuola*, 8.

33 The notion of female education aimed at social ends rather than individual ones was not expressed exclusively by men. On 12 April 1866, Olimpia Pieroni, an elementary school teacher in the province of Lucca, said in a speech, "non si chiedeva alla donna istruita di salire in cattedra ad insegnare le scienze: una buona educazione doveva essere diretta semplicemente a farle adempiere con cura i suoi doveri di figlia, di moglie, di madre [...] Chi biasima l'educazione della donna fa torto a se stesso, dimostra che non ama, o non stima il benessere delle famiglie, conciosiacché di qualunque condizione ella sia è sempre primiera maestra dell'uomo" [they were not asking the educated woman to sit at the teacher's desk and teach the sciences. A good education was simply directed to making her fulfil her duties as daughter, wife, and mother ... Who criticizes women's education is not doing justice to himself, and shows no love and respect for the well-being of families, since regardless of her condition, she is always the primary teacher of men]: quoted in Porciani, ed., *Le donne a scuola*, 71.

34 Quoted in Giampaolo Perugi, *Educazione politica in Italia 1860–1900* (Turin: Loescher, 1978), 194.

35 Fogazzaro defends the solitary condition of his work within the Italian literary world, refuting thus Serao's definition of movement when referring to the "Cavalieri dello spirito." See Antonio Fogazzaro, *Sonatine bizzarre: Scritti disperse* (Rome: Edizioni Ripostes, 1992), 61–5.

36 Serao, "La virtù delle donne," *Il Fanfulla della Domenica* 7 (30 August 1885), 35.

37 Rowland Strong, "An Italian Authoress: Mlle. Serao Has Scored a Great Triumph in Paris," *New York Times*, 16 July 1899, 17.

Appendix

1 Founded in Milan in 1848 by Alessandro Lampugnani, its name changed in 1860 to *Giornale delle Famiglie. La Ricamatrice*, and in 1874 it was joined

with *Corriere delle Dame*, becoming *Corriere delle Dame. Giornale delle Famiglie. Ricamatrice.* Its last issue was published on 26 July 1875 when Lampugnani, gravely ill and in financial distress, sold his periodicals to Edoardo Sonzogno.

2 All articles are transcribed as they were published.

3 Patrizia Zambon has identified the writer and journalist Ippolito Nievo (1831–61) as the author behind the pseudonym N.N. See note 55 in chapter 2.

4 Mezentius: a cruel Etruscan King, according to Roman mythology.

5 A reference to the suffragist movement and Amelia Bloomer, an American advocate of women's rights.

6 A journalist and politician (1813–93). He wrote for *Favilla* and *Perseveranza*, among other newspapers.

7 Daniele Manin, a patriot of the Risorgimento movement and president of the short-lived Republic of Venice (1848–9).

8 Joseph de Maistre (1753–1821). A native of Chambéry in Savoy, a northern province of the Kingdom of Piedmont-Sardinia, he was a reactionary writer known for his opposition to the Enlightenment and the French Revolution.

9 A whip, used in Russia for the corporal punishment of criminals.

10 Giulio Quaglia (1668–1751), a painter.

11 Born Gabriele Condulmer, he was pope from 1431 to 1447.

12 A daily paper founded in 1864 in Turin by Vittorio Bersezio. Initially named *Gazzetta piemontese*, it became *La Stampa* in 1895.

13 A weekly illustrated periodical, founded in 1873 by Emilio Treves. Many of the prominent and emerging literary figures of the time contributed to it, among them Ada Negri, Giosuè Carducci, Edmondo De Amicis, Neera, Matilde Serao, Giovanni Verga, and Luigi Pirandello.

14 Jules Michelet (1798–1874), a French historian and the author of *La Femme* (1860).

15 Elected to the Italian Parliament in1867, Salvatore Morelli fought many battles in favour of women's legal rights and emancipation.

16 Mount Taygetus, in Southern Greece. Marchesa Colombi is referring to a legend, popularized by Plutarch's writings on Sparta, according to which in Sparta deformed children were thrown off the cliffs of Mount Taygetus.

17 Synonym for "clerks." From Vittorio Bersezio's play *Le miserie del Signor Travetti* (1871), originally written in Piedmontese dialect (1863), which narrates the struggles of a government clerk, Ignazio Travet, who hopes to receive a raise in salary.

18 A short story about a young peasant boy who, after living for a little while in the city, decides to move back to the country. It is included in Neera's collection of short stories, *Iride* (1881).

19 A daily founded by Giuseppe Garibaldi on 21 April 1879. Mazzinians and members of the left contributed to it, among them Agostino Bertani, Alberto Mario, Adriano Saffi, and Antonio Fratti.

20 Serao's collection of short stories, among which were "Votazione femminile" and "Idillio di Pulcinella" (1879).

21 Maria Teresa of Savoy (1749–92), remembered here for her loyalty to Queen Marie Antoinette and her devotion to the members of the French royal family, for whom she died.

22 Italian translation of Emile Zola's *L'Assomoir* (1877).

23 Emile Zola's novel, which was translated into Italian in 1880.

24 Pseudonym of Olindo Guerrini (1845–1916), whose *Postuma* (1877), a collection of poems, was falsely attributed to a cousin, Lorenzo Stecchetti, who died of consumption at the age of thirty. The poems enjoyed great commercial success.

25 On 27 April 1880, Martino Cafiero published in *Corriere del Mattino* an article titled "La politica del cuore. Alla signorina Serao" in which he discussed both Agostino Bertani's criticism and Matilde Serao's defence of the monarchy. See page 250, note 8.

Bibliography

Albertocchi, Giovanni. *Sull'epistolario di Alessandro Manzoni: Disagi e malesseri di un mittente*. Florence: Cadmo, 1997.

Altman, Janet G. *Epistolarity: Approaches to a Form*. Columbus: Ohio State UP, 1982.

Altman, Janet G. "The Letter Book as a Literary Institution 1539–1789: Toward a Cultural History of Published Correspondences in France." *Yale French Studies* 71, 71 (1986): 17–62. http://dx.doi.org/10.2307/2930021.

Anderson, Benedict. *Imagined Communities*. London: Verso, 1991.

Antonelli, Giuseppe, Massimo Palermo, Danilo Poggiogalli, and Lucia Raffaelli, eds. *La scrittura epistolare nell'Ottocento*. Ravenna: Giorgio Pozzi Editore, 2009.

Artusi, Pellegrino. *La scienza in cucina e l'arte di mangiar bene*. Ed. Piero Camporesi. Turin: Einaudi, 2001.

Artusi, Pellegrino. *Science in the Kitchen and the Art of Eating Well*. Trans. Murtha Baca and Stephen Sartarelli. Toronto: U of Toronto P, 2003.

Banti, Alberto Mario. "Paura, dolore e lutto nel nazional-patriottismo ottocentesco." In *Politica ed emozioni nella storia d'Italia dal 1848 ad oggi*. Ed. Penelope Morris, Francesco Ricatti, and Mark Seymour. Rome: Viella, 2012: 43–51.

Barbèra, Gaspero. *Memorie di un editore*. Florence: Barbèra, 1883.

Barbiera, Raffaello. *Ideali e caratteri dell'Ottocento*. Milan: Treves, 1926.

Beebee, Thomas. *Epistolary Fiction in Europe: 1500–1850*. Cambridge: Cambridge UP, 1999.

Beebee, Thomas. "Publicity, Privacy, and the Power of Fiction in the Gunning Letters." *Eighteenth-Century Fiction* 20, 1 (Fall 2007): 61–88. http://dx.doi.org/10.1353/ecf.2008.0024.

Bell, Michael. *Sentimentalism, Ethics and the Culture of Feeling*. Houndmills, Basingstoke; and New York: Palgrave, 2000. http://dx.doi.org/10.1057/9780230595507.

Benjamin, Walter. *The Work of Art in the Age of Its Technological Reproducibility and Other Writings on Media*. Cambridge, MA: Harvard UP, 2008.

Bertoni Jovine, Dina. *Breve storia della scuola italiana*. Rome: Editori Riuniti, 1961.

Bertoni Jovine, Dina. *Storia dell'educazione popolare in Italia*. Bari: Laterza, 1965.

Bianchetti, Giuseppe. *Dei lettori e dei parlatori. Saggi due*. Rome: Vecchiarelli, 1989.

Bianchini, Angela. *Il romanzo d'appendice* Turin: ERI Edizioni Radio Televisione Italiana, 1969.

Bigazzi, Roberto. *I colori del vero: Vent'anni di narrativa 1860–1880*. Pisa: Nistri-Lischi, 1969.

Billiani, Francesca. "Delusional Identities: The Politics of the Italian Gothic and Fantastic in Igino Ugo Tarchetti's Trilogy *Amore nell'Arte* and Luigi Gualdo's Short Stories 'Allucinazione,' 'La Canzone di Weber,' and 'Narcisa.'" *Forum for Modern Language Studies* 4 (2008): 480–99.

Boella, Laura. *Sentire l'altro. Conoscere e praticare l'empatia*. Milan: Raffaello Cortina Editore, 2006.

Bonetta, Gaetano. "L'educazione sessuale della donna tra Otto e Novecento." In *E l'uomo educò la donna*. Ed. Carmela Covato and Maria Cristina Leuzzi. Rome: Editori Riuniti, 1989: 46–82.

Bonghi, Ruggero. *Perché la letteratura non sia popolare in Italia*. Ed. Edoardo Villa. Milan: Marzorati Editore, 1971.

Botteri, Inge. *Galateo e Galatei: La creanza e l'instituzione della società nella trattatistica italiana tra antico regime e stato liberale*. Rome: Bulzoni Editore, 1999.

Bourdieu, Pierre. *The Field of Cultural Production: Essays on Art and Literature*. Ed. Randal Johnson. New York: Columbia UP, 1993.

Brevini, Franco. *La letteratura degli italiani. Perché molti la celebrano e pochi la amano*. Milan: Feltrinelli, 2010.

Brilli, Antonio. *Il viaggiatore immaginario. L'Italia degli itinerari perduti*. Bologna: Il Mulino, 1997.

Buck-Morss, Susan. "The Flaneur, the Sandwichman and the Whore: The Politics of Loitering." *New German Critique, NGC* 39, 39 (Autumn 1986): 99–140. http://dx.doi.org/10.2307/488122.

Bulzoni, Lina, and Marcella Tedeschi. *Dalla scapigliatura al verismo*. Rome-Bari: Laterza, 1990.

Cadioli, Alberto. "Autori e pubblico." In *Manuale di letteratura italiana: Storia per generi e problemi*. Vol. IV. Ed Franco Brioschi and Constanzo di Girolamo. Turin: Bollati Boringhieri, 1993–6: 116–40.

Cadioli, Alberto, *La storia finta: Il romanzo e i suoi lettori nei dibattiti di primo Ottocento*. Milan: Il Saggiatore, 2001.

Caesar, Ann H. "Women Readers and the Novel in Nineteenth-Century Italy." *Italian Studies* 56 (2001): 80–97.

Caesar, Ann H., Gabriella Romani, and Jennifer Burns, eds. *The Printed Media in Fiu-de-siècle Italy: Publishers, Writers, and Readers*. Oxford: Legenda, 2011.

Campbell, Colin. *The Romantic Ethic and the Spirit of Modern Consumerism*. Oxford: Basil Blackwell, 1987.

Capuana, Luigi. *Lettere alla Assente*. Turin: Roux & Viarengo, 1904.

Capuana, Luigi. *Letteratura femminile*. Ed. Giovanna Finocchiaro Chimini. Catania: C.U.E.C.M., 1988.

Capuana, Luigi. "Rassegna letteraria: Storia di una capinera." In Carlo A. Madrignani, *Capuana e il naturalismo*. Rome-Bari: Laterza, 1970: 302–11.

Carnazzi, Giulio. "Verga e i veristi." In *L'Ottocento*. Ed. Armando Balduino. Padua: Piccin Nuova Libreria, 1997.

Carpi, Umberto. *Letteratura e società nella Toscana del Risorgimento*. Bari: De Donato Editore, 1974.

Causa, Cesare. *Il segretario galante ossia modo di scrivere lettere amorose sopra ogni sorta di argomento aggiuntovi l'epistolario degli amanti celebri ed altri scritti galanti*. Florence: Adriano Salani, 1890.

Cereto, Laura. *Collected Letters of a Renaissance Feminist*. Ed. Diana Robin. Chicago: U of Chicago P, 1997.

Chandler, James. "The Languages of Sentiment." *Textual Practice* 22, 1 (2008): 21–39.

Chartier, Roger. "Introduction: An Ordinary Kind of Writing." In *Correspondence: Models of Letter Writing from the Middle Ages to the Nineteenth Century*. Ed. Roger Chartier, Alain Boureau, and Cécile Dauphin. Princeton: Princeton UP, 1997.

Chartier Roger, and Guglielmo Cavallo, eds. *A History of Reading in the West*. Amherst and Boston: U of Massachusetts P, 1999.

Chemello, Adriana. *Alla lettera: Teorie e pratiche epistolari dai Greci al Novecento*. Padua: Guerini Studio, 1998.

Cook, Elizabeth H. *Epistolary Bodies: Gender and Genre in the Eighteenth-Century Republic of Letters*. Stanford, CA: Stanford UP, 1996.

Covato, Carmela. *Sap ere e pregiudizio: L'educazione delle donne tra '700 e '800*. Rome: Archivio Guido Izzi, 1991.

Croce, Benedetto. *Estetica come scienza dell'espressione e linguistica generale*. Bari: Laterza, 1909.

Croce, Benedetto. *La letteratura della Nuova Italia*. Bari: Laterza, 1915.

Cvetkovich, Ann. *Mixed Feelings: Feminism, Mass Culture, and Victorian Sensationalism*. New Brunswick, NJ: Rutgers UP, 1992.

Darnton, Robert. "What Is the History of Books?" In *The Book History Reader*. Ed. David Finkelstein and Alistair McCleery. London: Routledge, 2006: 9–26.

Dauphin, Cécil. "Letter-Writing Manuals in the Nineteenth Century." In *Correspondence: Models of Letter Writing from the Middle Ages to the Nineteenth Century*. Ed. Roger Chartier, Alain Boureau, and Cécile Dauphin. Princeton: Princeton UP, 1997: 112–57.

De Amicis, Edmondo. *Fra scuola e casa: Bozzetti e racconti*. Milan: Treves, 1906.

De Gubernatis, Angelo. *F. Dall'Ongaro. Il suo epistolario scelto: Ricordi e Spogli*. Florence: Tipografia Editrice dell'Associazione, 1875.

Della Coletta, Cristina. *World's Fairs Italian Style: The Great Exhibitions in Turin and Their Narratives, 1860–1915*. Toronto: U of Toronto P, 2006.

Denby, David. *Sentimental Narrative and the Social Order in France*. Cambridge and New York: Cambridge UP 1999.

De Nunzio Schilardi, Wanda. *L'invenzione del reale: Studi su Matilde Serao*. Naples: Palomar, 2004.

De Nunzio Schilardi, Wanda, *"La Settimana" di Matilde Serao*. Pisa: Giardini Editori e Stampatori, 2006.

De Roberto, Federico, *L'albero della scienza*. Caltanissetta: Edizioni Lussografica, 1997.

De Roberto, Federico, *L'amore: Fisiologia, psicologia, morale*. Milan: Casa Editrice Galli, 1895.

De Roberto, Federico. *Casa Verga e altri saggi verghiani*. Ed. Carmelo Musumarra. Florence: Le Monnier, 1964.

De Sanctis, Francesco. *Il pensiero educativo* Florence: Vallecchi Editore, 1937.

Dickie, John. *Delizia! The Epic History of the Italians and Their Food*. New York: Free Press, 2008.

Dierks, Konstantin. *In My Power: Letter Writing and Communication in Early America*. Philadelphia: U of Pennsylvania P, 2009.

di Fazio, Margherita. *La lettera e il romanzo: Esempi di comunicazione epistolare nella narrativa*. Rome: Nuova Amica Editrice, 1996.

Dixon, Thomas. *From Passions to Emotions: The Creation of a Secular Psychological Category*. Cambridge: Cambridge UP, 2003. http://dx.doi.org/10.1017/CBO9780511490514.

Doglio, Maria Luisa. *Lettera e Donna. Scrittura epistolare al femminile tra Quattro e Cinquecento*. Rome: Bulzoni, 1993.

Eco, Umberto. *The Role of the Reader*. Bloomington and London: Indiana UP, 1979.

Evans, Dylan. *Emotion: The Science of Sentiment*. Oxford: Oxford UP, 2002.

Faccioli, Emilio. "Un editore popolare di orientamento moderato: Adriano Salani." In *Editori a Firenze nel secondo Ottocento*. Atti del Convegno (13–15 novembre 1981). Ed. Ilaria Porciani. Florence: Leo S. Olschki Editore, 1983: 367–80.

Favret, Mary. *Romantic Correspondence: Women, Politics, and the Fiction of Letters.* Cambridge: Cambridge UP, 1993.

Ferrucci, Caterina Franchini. *Della educazione della donna italiana.* Turin: Unione Tipografico-Editrice, 1855.

Fiducia, Saverio. *Passeggiate sentimentali: Catania di ieri e di oggi.* Catania: Giannotta, 1966.

Finotti, Fabio. "Ritratto, maschera, fisionomia. Il genere epistolare e il carteggio Croce- Prezzolini." *Lettere Italiane* 43, 1 (1991): 88–104.

Foscolo, Ugo. *Storia della letteratura italiana per saggi.* Ed. Mario Alighiero Manacorda. Turin: Einaudi, 1979.

Foucault, Michel. *The Archeology of Knowledge & the Discourses on Language.* New York: Pantheon Books, 1972.

Foucault, Michel. *History of Sexuality. An Introduction.* Vol. I. New York: Vintage Books, 1990.

Franchini, Silvia. *Editori, lettrici e stampa di moda.* Milan: FrancoAngeli, 2002.

Franzina, Emilio. "L'epistolografia popolare e i suoi usi." *Materiali di Lavoro* 1–2 (1987): 21–63.

Friedberg, Anne. *The Virtual Window from Alberti to Microsoft.* Cambridge, MA: Massachusetts Institute of Technology P, 2006.

Gabelli, Aristide. "L'Italia e l'istruzione femminile." *Nuova Antologia* 15, 9 (September 1870): 145–67.

Gambacorti, Irene. *Verga a Firenze: Nel laboratorio di Storia di una capinera.* Florence: Le lettere, 1994.

Gambarota, Guglielmo. *Inchiesta sulla donna.* Florence and Turin: Fratelli Bocca, 1899.

Gelli, Jacopo. *Come devo scrivere le mie lettere? Esempi di lettere e di scritture private per tutte le occasioni della vita.* Milan: Ulrico Hoepli, 1900.

Gherarducci, Isabella, and Enrico Ghidetti. *Guida alla lettura di Verga.* Florence: La Nuova Italia, 1994.

Ghidetti, Enrico. *Verga: Guida storico-critica.* Rome: Editori Riuniti, 1979.

Gibelli, Antonio, and Fabio Caffarena. "Le lettere degli emigranti." In *Storia dell'emigrazione italiana.* Ed. Piero Bevilacqua, Andreina De Clementi, and Emilio Franzina. Rome: Donzelli, 2001: 563–74.

Gilroy, Amanda, and W.M. Verhoeven, eds. *Epistolary Histories: Letters, Fiction, Culture.* Charlottesville: UP of Virginia, 2000.

Gisolfi, Anthony M. *The Essential Matilde Serao.* New York: Las Americas Publishing Company, 1968.

Goldsmith, Elizabeth C., ed. *Writing the Female Voice: Essays on Epistolary Literature.* Boston: Northwestern UP, 1989.

Habermas, Jürgen. *The Structural Transformation of the Public Sphere*. Cambridge, MA: MIT P, 1991.

Harding Jennifer, and E. Deidre Pribram, *Emotions: A Cultural Studies Reader*. London and New York: Routledge, 2009.

Helstosky, Carol. "Recipe for the Nation: Reading Italian History through La Scienza in Cucina and La Cucina Futurista." *Food & Foodways* 11, 2–3 (2003): 113–40. http://dx.doi.org/10.1080/07409710390242372.

Hendler, Glenn. *Public Sentiments: Structures of Feeling in Nineteenth-Century American Literature*. Chapel Hill: U of North Carolina P, 2001.

Henkin, David. *The Postal Age: The Emergence of Modern Communications in Nineteenth-Century America*. Chicago: U of Chicago P, 2006.

Herget, Winfried, ed. *Sentimentality in Modern Literature and Popular Culture*. Tübingen: Gunter Narr Verlag, 1991.

How, James. *Epistolary Spaces: English Letter Writing from the Foundation of the Post Office to Richardson's Clarissa*. Aldershot, Hants; Burlington, VT: Ashgate, 2003.

Kauffman, Linda. *Special Delivery*. Chicago: U of Chicago P, 1989.

Keen, Suzanne. *Empathy and the Novel*. Oxford: Oxford UP, 2007. http://dx.doi.org/10.1093/acprof:oso/9780195175769.001.0001.

Lolla, Maria Grazia. "Reader/Power: The Politics and Poetics of Reading in Post-Unification Italy." In *The Printed Media in Fin-de-Siècle Italy: Publishers, Writers, and Readers*. Ed. Ann H. Caesar, Gabriella Romani, and Jennifer Burns. Oxford: Legenda, 2011: 20–37.

Lombardo, Patrizia. "Introduction: The Intelligence of the Heart." *Critical Quarterly* 50, 4 (December 2008): 1–11. http://dx.doi.org/10.1111/j.1467-8705.2008.00857.x.

Luzzi, Joseph. "Verga *Economicous*: Language, Money, and Identity in *I Malavoglia* and *Mastro-don Gesualdo*." In *The Printed Media in Fin-de-Siècle Italy: Publishers, Writers, and Readers*. Ed. Ann H. Caesar, Gabriella Romani, and Jennifer Burns. Oxford: Legenda, 2011: 39–48.

Mantegazza, Paolo. *Fisiologia della donna*. Vol 2. Milan: Treves, 1883.

Manzoni, Alessandro. *Edizione nazionale delle opere di Alessandro Manzoni. Carteggio, Alessandro Manzoni, Claude Fauriel*. Vol. 27. Ed. Irene Botta. Milan: Centro Nazionale Studi Manzoniani, 2000.

Manzoni, Alessandro. *Tutte le opere*. Ed. Mario Martelli. Florence: Sansoni Editore, 1973.

Marchesa Colombi [pseud. of Marie Antoinette Torriani]. *Gente per bene*. Novara: Interlinea, 2000.

Messina, Maria. *Piccoli gorghi*. Palermo: Sellerio editore, 1997.

Mestica, Giovanni. *Istituzioni di letteratura*. Florence: G. Barbera Editore, 1882.

Milne, Esther. *Letters, Postcards, Emails: Technologies of Presence*. New York: Routledge, 2010.

Morandini, Giuliana. "Introduzione." Introduction to Marchesa Colombi, *Un matrimonio in provincia*. Novara: Interlinea, 1999: 5–12.

Moretti, Franco. *Atlas of the European Novel 1800–1900*. London: Verso, 1999.

Moretti, Franco. *Signs Taken for Wonders. Essays in the Sociology of Literary Forms*. London: Verso, 1983. http://dx.doi.org/10.2307/1772338.

Mori, Maria Teresa. *La sociabilità delle elite nell'Italia dell'Ottocento*. Rome: Carocci, 2000.

Morpugo, Emilio. *La posta e la vita sociale*. Turin, Rome, Florence: Ermanno Loescher, 1883.

Murialdi, Paolo. *Storia del giornalismo italiano*. Bologna: Il Mulino, 2006.

Musumarra, Carmelo. *Verga minore*. Pisa: Nistri-Lischi, 1965.

Neera [pseud. of Anna Radius Zuccari]. *Una giovinezza del secolo XIX*. Milan: Feltrinelli, 1980.

Nelson, Elizabeth White. *Market Sentiments: Middle-Class Market Culture in Nineteenth-Century America*. Washington, DC: Smithsonian Books, 2004.

Nevers, Emilia. *Galateo della Borghesia: Norme per trattar bene*. Turin: Giornale delle donne, 1883.

Nievo, Ippolito. *Scritti giornalistici alle lettrici*. Ed. Patrizia Zambon. Lanciano, Chieti: Casa Editrice Rocco Carabba, 2008.

Nussbaum, Martha C. *Poetic Justice: The Literary Imagination and Public Life*. Boston: Beacon Press, 1995.

Ojetti, Ugo. *Alla scoperta dei letterati*. Milan: Fratelli Dumolard Editori, 1895.

Palermo, Antonio. "Le due narrative di Matilde Serao." *Da Mastriani a Viviani: Per una storia della letteratura a Napoli fra Ottocento e Novecento*. Naples: Liguori Editore, 1972.

Paoloni, Giovanni, ed. *Le poste in Italia*. Vols I and II. Bari: Laterza, 2005.

Pautasso, Sergio. "Introduzione." Introduction to Giovanni Verga, *Storia di una capinera* (Milan: Arnoldo Mondadori, 1991: v–xiii.

Petrucci, Armando. *Scrivere lettere: Una storia plurimillenaria*. Rome-Bari, Laterza, 2008.

Picci, Giuseppe. *Guida allo studio delle belle lettere e al comporre con un manuale dello stile epistolare*. Milan: G. Gnocchi Editore, 1855.

Porciani, Ilaria, ed. *Le donne a scuola*, Florence: Tipografia il Sedicesimo, 1987.

Porciani, Ilaria, ed. *Famiglia e nazione nel lungo Ottocento italiano: Modelli, strategie, reti di relazioni*. Rome: Viella, 2006.

Prima relazione sul servizio postale in Italia. Anno 1863. Turin: Tipografia Fodratti, 1864.

Pulcini, Elena. "J.J. Rousseau: L'immaginario e la morale." Introduction to Jean-Jacques Rousseau, *Giulia o la nuova Eloisa*. Milan: Rizzoli, 1992: iii–lxxx.

Pulcini, Elena. "Per una breve storia delle emozioni." In *Il teatro delle passioni: Ragione e sentimento nell'età moderna*. Ed. Nadia Boccara and Letizia Gai. Viterbo: Sette Città, 2003: 169–83.

Pupino, Angelo. *Notizie dal reame*. Naples: Liguori, 2004.

Racevskis, Roland. "Dynamics of Time and Postal Communication in Mme de Sévigné's Letters." *Papers on French Seventeenth-Century Literature* 26, 50 (1999): 51–9.

Ragone, Giovanni. *Un secolo di libri: Storia dell'editoria in Italia dall'unità al post-moderno*. Turin: Einaudi, 1999.

Raicich, Marino. *Scuola, cultura e politica da De Sanctis a Gentile*. Pisa: Nistri-Lischi, 1981.

Ray, Meredith K. *Writing Gender in Women's Letter Collections of the Italian Renaissance*. Toronto: U of Toronto P, 2009.

Raya, Gino. *Carteggio Verga-Capuana*. Rome: Edizioni dell'Ateneo, 1984.

Raya, Gino. *Verga e i Treves*. Rome: Herder Editore, 1986.

Raya, Gino. *Vita di Giovanni Verga*. Rome: Herder Editore, 1990.

Riall, Lucy. *Garibaldi: Invention of a Hero*. New Haven: Yale UP, 2007.

Romagnoli, Sergio. "Un traguardo editoriale: La Carducciana." In *Editori a Firenze nel secondo Ottocento*. Ed. Ilaria Porciani. Florence: Leo Olschki Editore, 1983.

Salsini, Laura. *Addressing the Letter: Italian Women Writers' Epistolary Fiction*. Toronto: U of Toronto P, 2010.

Salsini, Laura. *Gendered Genres: Female Experiences and Narrative Patterns in the Works of Matilde Serao*. Madison, NJ: Fairleigh Dickinson UP, 1999.

Sassatelli, Roberta. *Consumer Culture: History, Theory and Politics*. London: Sage, 2007.

Scarfoglio, Edoardo. *Il libro di Don Chisciotte*. Ed. Carlo Alberto Madrignani. Naples: Liguori, 1990.

Scarpellini, Emanuela. *L'Italia dei consumi dalla Belle Époque al nuovo millennio*. Rome-Bari: Editori Laterza, 2008.

Serao, Matilde. *Gli amanti*. Naples: Francesco Perrella, 1908.

Serao, Matilde. "I cavalieri dello spirito." In Antonio Fogazzaro. *Sonatine bizzarre: Scritti dispersi*. Rome: Ripostes, 1992: 53–9.

Serao, Matilde. *Fior di passione*. Milan: Casa Editrice Baldini, Castoldi & C., 1899.

Serao, Matilde. *Il giornale*. Naples: Francesco Perrella Editore, 1906.

Serao, Matilde. *Lettere d'amore*. Catania: Cav. Niccolò Giannotta Editore, 1901.

Serao, Matilde. *Novelle sentimentali*. Livorno: S. Belforte E C., Editori, 1902.

Serao, Matilde. *Saper vivere: Norme di buona creanza*. Florence: Passigli Editori, 1989.

Soldani, Simonetta. "Salotti dell'Ottocento: Qualche riflessione." In *Salotti e ruolo femminile in Italia: Tra fine Seicento e primo Novecento*. Ed. Maria Luisa Berri and Elena Brambilla. Venice: Marsilio, 2004: 553–68.

Spera, Francesco. "Amore, follia e morte nella storia dell'artista." In *Igino Ugo Tarchetti e la Scapigliatura*. Atti del Convegno. Ed. Franco Contorbia. San Salvatore Monferrato: Comune di San Salvatore Monferrato, 1976.

Spinazzola, Vittorio. *Letteratura e popolo borghese*. Milan: Unicopli, 2000.

Spirito, Ugo, ed. *Il pensiero pedagogico del positivismo*. Florence: Sansoni, 1956.

Stewart-Steinberg, Suzanne. *The Pinocchio Effect: On Making Italians, 1860–1920*. Chicago: U of Chicago P, 2007.

Tarchetti, Iginio Ugo. *Amore nell'arte*. Turin: Lindau, 1993.

Tellini, Gino, ed. *Scrivere lettere: Tipologie epistolari nell'Ottocento italiano*. Rome: Bulzoni, 2002.

Tenca, Carlo. *Giornalismo e letteratura nell'Ottocento*. Rocca San Casciano: Cappelli Editore, 1959.

Tobia, Bruno. "Una cultura per la nuova Italia." In *Storia d'Italia*. Vol. 2. Ed. Giovanni Sabbatucci and Vittorio Vidotto. Bari: Laterza, 1995: 427–529.

Tortorella Esposito, Giovanna. "Tra reale ed immaginario: La dimensione dell'artista nella trilogia 'Amore dell'arte' di I.U. Tarchetti." *Esperienze Letterarie* 19, 3 (1994): 51–66.

Tortorelli, Gianfranco. "I libri più letti dal popolo italiano: Un'inchiesta del 1906." *Studi di Storia dell'Editoria Italiana*. Bologna: Patron editore, 1989: 153–69.

Trasciatti, Mary Anne. "Letter Writing in an Italian Immigrant Community: A Transatlantic Tradition." *Rhetoric Society Quarterly* 39, 1 (2009): 73–94. http://dx.doi.org/10.1080/02773940802561884.

Trotta, Donatella. *La via della penna e dell'ago: Matilde Serao tra giornalismo e letteratura*. Naples: Liguori Editore, 2008.

Turnaturi, Gabriella. *Betrayals*. Chicago: U of Chicago P, 2007.

Turnaturi, Gabriella. *Signore e signori d'Italia. Una storia delle buone maniere*. Milan: Feltrinelli, 2011.

Valesio, Paolo. *Novantiqua: Rhetoric as a Contemporary Theory*. Bloomington: Indiana UP, 1980.

Verga, Giovanni . *Lettere sparse*. Ed. Giovanna Finocchiaro Chimirri. Rome: Bulzoni, 1979.

Verga, Giovanni. *Sparrow*. Trans. Christine Donougher. Sawtry and New York: Dedalus/Hippocrene, 1994.

Verga, Giovanni. *Storia di una capinera*. Ed. Maura Brusadin. Pordenone: Edizione Studi Tesi, 1985.

Watt. Ian. *The Rise of the Novel*. Berkeley and Los Angeles: U of California P, 1957.

Williams, Simon. "Modernity and the Emotions: Corporeal Reflections on the (Ir)Rational." In *Emotions: A Cultural Studies Reader*. Ed. Jennifer Harding and E. Deidre Pribram. London and New York: Routledge, 2009: 139–56.

Zarri, Gabriella, ed. *Per lettera: La scrittura epistolare femminile tra archivio e tipografia, secoli XV–XVII*. Rome: Viella, 1999.

Index

www.ingramcontent.com/pod-product-compliance
Lightning Source LLC
LaVergne TN
LVHW040151080826
844660LV00014B/917/J

* 9 7 8 1 4 4 2 6 4 7 0 8 4 *